AF443411

C 15 0035933

WATERBORNE COATINGS

Emulsion and Water-Soluble Paints

Other Books by Charles R. Martens

Alkyd Resins

Emulsion and Water-Soluble Paints and Coatings

Technology of Paints, Varnishes and Lacquers (Editor)

WATERBORNE COATINGS

Emulsion and Water-Soluble Paints

Charles R. Martens
Consultant

VNR
VAN NOSTRAND REINHOLD COMPANY
NEW YORK CINCINNATI ATLANTA DALLAS SAN FRANCISCO
LONDON TORONTO MELBOURNE

Van Nostrand Reinhold Company Regional Offices:
New York Cincinnati Atlanta Dallas San Francisco

Van Nostrand Reinhold Company International Offices:
London Toronto Melbourne

Library of Congress Catalog Card Number: 80-17143
ISBN: 0-442-25137-8

Manufactured in the United States of America

Published by Van Nostrand Reinhold Company
135 West 50th Street, New York, N.Y. 10020

Published simultaneously in Canada by Van Nostrand Reinhold Ltd.

15 14 13 12 11 10 9 8 7 6 5 4 3 2 1

Library of Congress Cataloging in Publication Data

Martens, Charles R.
Waterborne coatings.

Includes bibliographies and index.
1. Emulsion paint. I. Title.
TP935.M29 667'.63 80-17143
ISBN 0-442-25137-8

Preface

Waterborne Coatings: Emulsion and Water-Soluble Paints covers the history, development, formulation, and uses of waterborne paints. Since my first book on this subject, *Emulsion and Water-Soluble Paints and Coatings,* was published in 1964, much progress has been made and many new developments have occurred.

I wish to express my appreciation to the many companies, particularly raw material suppliers, that provided data, information, and photographs.

CHARLES R. MARTENS

Contents

WATERBORNE COATINGS

Emulsion and Water-Soluble Paints

1
Introduction

The most significant development in the protective coating field since World War II has been the rapid growth of waterborne paints. First waterborne paints gained popularity in trade sales paints because of their minimal odor and ease of cleanup; later waterborne paints were used in industrial applications because of the reduced fire hazard and reduction in the amount of solvent released into the environment. In many cases, the waterborne paints have properties superior to those of the solvent paints they replace.

TERMINOLOGY

The term "waterborne" is the correct generic term for coatings containing some water in the volatiles. The term "water-based coating" is often used to designate a waterborne coating. This is a misnomer in that the base of the coating is normally the binder (polymer, resin, etc.). Calling a coating water-based is equivalent to calling a coating toluene-based or mineral-spirits-based. Normally, a coating would be termed alkyd-based or acrylic-based because of the polymer composition. Thus, the proper terminology for an acrylic coating that has water as the volatile component would be an "acrylic-based waterborne coating" or simply an "acrylic waterborne coating." It is recommended that the term "water-based" be dropped.

Waterborne coating includes aqueous coating, water-thinned, water-reducible, emulsion, latex, casein, cement, and silicate paints, all of which contain some water when applied.

In the simplest form, paints consist of pigments, binders, and volatiles. In a waterborne system, the pigments are similar to those used in a solvent system in most respects. The binder is usually an organic polymer, as, for example, an acrylic. The binder may be in solution or dispersion form. The volatile is water alone or with a miscible cosolvent such as an alkyd, glycol-ether, and so forth.

HISTORY

Waterborne paints were the earliest types of paints; they may be traced to the dawn of history. The caveman first expressed his artistic abilities by daubing colored mud on the walls of his cave. Then crushed berries, blood, eggs, milk, dandelion, milkweed, or tree sap and other materials were added to improve adhesion and durability. Water was the only solvent available at that time.

The Egyptians decorated their walls with paints executed in distemper during the years 3000 to 2000 B.C. The great period of wall painting was in the 12th to 10th Dynasty, which fixes the date before 1600 B.C. The term "distemper" is of British origin and refers to waterborne materials containing glue, casein, egg whites, and vegetable gums, such as gum arabic and recently dextrinized starches. Its use has continued to the present day in artists' watercolors and poster showcard colors. The frescos of the Romans and early Italians were applications of pigments on uncured plaster where the colors became fixed by carbonization and drying of the lime. The ancient Hebrews used milk or curd paints for decorating the home.

Whitewash, which was essentially water-slaked lime, was used during colonial days and in the early history of the United States. Tom Sawyer's whitewashed fence is described in fiction. Various materials were added to the lime to improve the coating. Portland cement was incorporated to improve the durability of seashore applications, a usage that gradually evolved into present-day cement paints. Skim milk and later casein were added to improve adhesion and durability; and the additions of pigments, whiting, and clay, and finally the replacement of the lime traced the evolution from whitewash to casein paints. Drying oils were added to casein paints to improve water resistance. The first U.S. paint patent was issued in 1865 (U.S. Patent 50,058) and covers a composition based on zinc oxide, potassium hydroxide, resin, milk, and linseed oil.

Casein paints were continually improved so that by the 1930s they contained high hiding pigments (high refractive index) similar to those used in oil paints. These casein paints were marketed in paste form as well as in powder form. In order to market a paste-form paint, it is necessary to stabilize the casein against hydrolysis and putrefaction. The outstanding feature of casein paints is their low cost. These paints are easy to apply and have good one-coat hiding. However, casein paints are porous and thus soil easily. Although casein paints will withstand some washing, it is difficult to remove stains.

The addition of a drying oil to a casein paint produced a tighter film and improved dirt resistance. At the Chicago Century of Progress Exposition in 1933, casein paints reinforced with tung oil were used to produce bright colors in a flat sheen paint. As the amount of oil increased, the evolution from a casein to an emulsion paint took place. When the amount of oil exceeded a 2:1 ratio, a true emulsion paint was attained. In many cases the oil was replaced with an alkyd to give a more rapid-drying paint. Resin emulsion paints were used for painting exterior surfaces at the New York World's Fair and the Golden Gate Exposition with good results. The casein in emulsion paints then became a thickener and emulsion stabilizer, and the oil or alkyd produced a tight film with improved resistance to staining. Limed linseed oil under such names as "Calcicoater" could contain limited amounts of water in the dispersed phase for use on interior walls. Emulsion oil paints enjoyed great popularity during World War II because the limited amounts of drying oils available for decorative painting were most effectively utilized in emulsion paints.

In the 1940s there were three major developments in waterborne coatings. In the United States the classic low-cost interior wall paint based on a protein medium was improved by the emulsification of a drying oil–resin combination, with the acidity high enough for partial neutralization with ammonia. This yielded a ready-to-use waterborne emulsion wall paint. Proteins and water-dispersible modified celluloses were still necessary minor components for controlling rheology. Fast drying required that the oleoresinous binder be highly polymerized so that early stages of drying were accomplished mainly by evaporation of water followed by insolubilization of the oil portion by oxygen cross-linking from the air using driers or metal catalysts to attain fair resistance to abrasion, washing, and so on. This fast, early drying required the substantial absence of the usual hydrocarbon solvents. The high viscosity of the vehicle made manufacture difficult because of pumping and straining problems.

Use of the millions of gallons of these paints produced in the 1940s was promoted by the evolution of the hand roller, which made application to wall surfaces by do-it-yourselfers very easy. Inherent limitations of the low-gloss films channeled their use to large surfaces such as walls or ceilings where minute surface irregularities were less obvious.

Camouflage paints prepared as pastes for dilution in the field were a very important development in World War II for waterborne coatings. These oil-in-water pastes were reduced in the field with one volume of gasoline or three volumes of water. The pastes had storage stability in temperature cycling from 5 to 70°C and generally had good exterior durability.

In Germany during and after World War II polyvinyl acetate and polyvinyl propionate latexes were used for exterior paints because of the scarcity of linseed oil. Their use pioneered the development of modern latex paints.

At the end of World War II the era of synthetic latex paint began in the United States. Government Rubber Styrene (GRS) had been used for military tires during the war because of the loss of natural rubber supplies after the attack on Pearl Harbor. GRS rubber has a composition of 60 parts of butadiene to 40 parts of styrene. Polybutadiene is the softer polymer. In styrene-butadiene latex for paints, the ratio was reversed: 40 parts of butadiene to 60 parts of styrene was used to produce a harder polymer.

Thus, in the 1950s, styrene-butadiene latexes were introduced for interior wall paints. Then, as commercial sources of polyvinyl acetate became available in the United States, that substance was used for trade sales paints; later, with the introduction of acrylic latexes, exterior latex house paints were developed.

In the early fifties waterborne alkyd paints were introduced for industrial applications in both water-soluble and water-dispersible forms. In 1953 a fire occurred at the Livonia, Michigan Transmission Plant of the General Motors Corporation. The fire, which was caused by welding sparks igniting the residue in a drip pan, spread rapidly to the tar roof and traveled the full length of the 1800-foot building in approximately 1¼ hours. As a result of this fire, an effort was made to reduce the amount of inflammable materials used. One way to do so was to use waterborne paints.

By 1970 semigloss latex paints were introduced. These paints required the development of smaller-particle-size latices along with better-coalescing solvents.

One development in waterborne industrial coatings was electrodeposition. Electrocoating on steel or aluminum of paints based on anionic or more recently cationic polymers has been a major advance in reducing solvent pollution in the atmospher. In this method, the metal being coated is the anode (or cathode), and the resin in water-soluble form with pigments is carried to it by an electric current. At the anode the coating particles are discharged by the current, which forces the liquid away from the deposited coating. The coating is thus converted from a water-soluble to a water-insoluble form. This method will deposit the coating in complex and inaccessible areas. Each tank so used is limited to one color, and it is a difficult procedure to change the contents of a tank.

Textile printing and padding colors based on water-in-oil emulsions and reversals thereof brought early technical advances to a branch of industrial coatings. Photographic emulsions are another area of sophisticated waterborne coatings.

The chronological history of the coating industry in the United States is shown in Table 1.1.

FUNCTION OF PAINT

The two primary functions of paint are decoration and protection. Decorative effects may be produced by color, gloss, or texture or a combination of these characteristics. The buyer who purchases paint for a living room wall is primarily interested in decoration. The paint should be durable and washable, but a room is usually painted to change its color and not because the paint has failed. A secondary decorative function is lighting. The color of the surface affects its reflectance. The proportion of light that is released by a surface is expressed as a percentage of complete reflectance. The following are approximate reflectance values for various colors:

TABLE 1.1. HISTORY OF THE COATING INDUSTRY IN THE UNITED STATES

1804	First white lead production in the United States
1815	First varnish production in the United States
1865	First U.S. patent (50,068) on water paint to D. P. Flinn
1867	Production of ready-mixed paints in the United States
1910	Casein powder paints available on market
1923	Nitrocellulose lacquers
1924	Titanium dioxide pigments introduced
1924	Modified phenolic resins commercially available
1927	Alkyd resins commercially available
1928	Oil-soluble phenolics
1929	Urea-formaldehyde resins for modification of alkyds
1930	Chlorinated rubber first produced
1930	Casein paste paints put on the market
1930	Molybdate orange introduced
1933	Vinyl copolymer—vinyl acetate–chloride introduced
1934	Oil-emulsion paints put on market
1937	Phthalocyanine blue pigment introduced
1939	Commercial production of dehydrated castor oil
1939	Melamine-formaldehyde resins available
1939	Ethyl cellulose introduced
1939	Polyurethane resins introduced
1944	Silicone resins commercially available
1948	Latex wall finishes based on styrene-butadiene put on market
1950	Epoxy resins commercially available
1950	Polyvinyl acetate and acrylic copolymer latices commercially available
1957	Latex house paint introduced
1962	Electrodeposition of water paints
1966	Rule 66 solvent regulations on paints
1968	Semigloss latex paints introduced
1972	Titanium dioxide water slurries commercially available

White	90–80%
Very light tints	80–70%
Light tints	70–60%
Medium to dark tints	60–20%
Deep colors	20–3%
Black	2–1%

Because there is a multiple reflection of light from one surface to another, the effect of room lighting is more pronounced than is indicated in the above figures.

The protective function of paints includes protection of surfaces from air, water, microorganisms, and chemicals (i.e., acids, alkalis, and atmospheric fumes), and improved mechanical properties on some materials through better hardness, abrasion resistance, and so on.

The protective coating may be the paint on a wooden boat, for protection against rotting; the coating on the interior metal lining of metal cans or drums, to prevent corrosion from food or chemicals; coatings on electrical parts, for excluding moisture; fire-retardant paints, to protect combustible surfaces; or coatings on porous surfaces such as concrete and plaster, for ease of cleaning.

TYPES OF PAINT

The protective coatings industry is divided into two broad categories. Trade sales or shelf goods include products sold to consumers, contractors, and professional painters for use on new construction or maintenance. This is on-site painting, and these paints are sometimes known as architectural paints. Shelf goods are usually air dry finishes. Industrial maintenance finishes, which are high-performance finishes particularly in regard to corrosion resistance, are usually included in this class.

The growth of waterborne paints in the trade sales market has been due to their ease of application, easy cleanup, and minimal odor. They now dominate the consumer market and are usually sold in the form of latex paints. When latex paints were introduced, alkyd paints dominated the architectural market. Latex paints first replaced alkyds in wall finishes. The new systems could not compete with alkyd paints, however, particularly in exterior house paints where the durable oil-based paints had good adhesion and weathering properties. Continued improvements in latex paints made them more competitive, more durable, and easy to apply; improved latex paints showed good color retention, wet adhesion, leveling, and flow.

As a result, waterborne paints have displaced alkyd and oil house paints in most regions. Only in the far northern areas of the United States are alkyd paints required, owing to the severe winter weather. Latex paints have also made significant inroads on alkyd gloss enamels which were used extensively in the interior-house-paint market.

Waterborne paints now account for roughly 90% of the gallonage of flat interior coatings. Their share of the semigloss interior market is over 50%. Their share of exterior paints is now about 70%.

Recently in some areas, government regulations have limited the amount of solvent or volatile organic compounds that can be used in some trade sales paints, so that waterborne paints are the only products that can meet these regulations.

Chemical coatings or industrial product finishes are produced to user specifications and sold to manufacturers for factory application to such items as appliances, automobiles, metal containers, house siding, and so on. The paint is applied by spray, dip, roller, or electrodeposition. The paint is usually applied on a production line and then baked.

Waterborne paints are used in industrial product finish lines because of the minimal fire hazard and because of government regulations restricting the release of organic solvents to the atmosphere. Conventional solvent paints may be used, but the solvent must not enter the atmosphere.

There are several possible ways to reduce solvent content of coatings:

Waterborne coatings
High-solids coatings
Powder coatings

Generally, waterborne coatings require minimum changes of the production line. In industrial waterborne coatings, water-soluble coatings are usually used.

Tables 1.2 and 1.3 list some advantages of using waterborne coatings in trade sales and industrial finishes.

TABLE 1.2. ADVANTAGES OF WATERBORNE COATINGS—TRADE SALES.

Properties	Advantages
Easy application	Easy brushing and roller coating
Minimal odor	Reduced odor when painting in the home
Cleanup with water	Easy cleanup if done properly
Paint moist surfaces	No delay waiting for surfaces to dry out

TABLE 1.3. ADVANTAGES OF WATERBORNE COATINGS—INDUSTRIAL.

Properties	Advantages
Low flammability	Increased worker safety, lower insurance rates.
Reduced toxicity, odor	Increased worker safety, compliance with occupational safety and health regulations.
Reduced environmental pollution	Healthier environment, compliance with clean air and water legislation
Cleanup with water	Easy cleanup if done properly
Use conventional application methods, spray, direct and reverse roll coater, dip electrostatic disc or cone	Save cost of new finishing lines
Holdout (emulsion)	Coat porous surfaces such as particle board.

AIR POLLUTION

Air pollution regulations limit the amount of organic solvents that can be discharged into the atmosphere. The term used for solvents is "volatile organic compounds" (VOC). A volatile organic compound is defined as any compound of carbon that has a vapor pressure greater than 0.1 millimeter of mercury at standard conditions (20°C—760 mm pressure mercury), but excluding methane, carbon dioxide, carbonic acid, metallic carbides or carbonates, 1,1,1 trichloroethane, and ammonium carbonate.

Hydrocarbons are detrimental to the atmosphere as they react with nitrogen oxides in the presence of sunlight to form oxidants, producing smog which irritates the eyes, nasal passages, and respiratory systems.

Volatile organic compounds are emitted from a variety of sources. Total emissions in the United States for the year 1975 were estimated by the Environmental Protection Agency (EPA) to be about 31 million tons, of which 19 million tons were from stationary sources. Evaporation of organic solvents contributed about 44% of the total amount from stationary sources.

Emissions of organic air pollutants can be reduced on paint application equipment by use of: (1) add-on control devices that either destroy or collect the organic solvents for reuse or disposal; or (2) process or material changes that reduce or eliminate the use of organic solvents.

Today the principal add-on control devices for the control of volatile organics are:

1. Catalytic and noncatalytic (thermal) incinerators
2. Activated carbon and other types of adsorbers
3. Liquid scrubbers or adsorbers
4. Condensers that use refrigeration or compressors

Incineration is the technique most universally applied by industry, but it usually requires supplemental fuel. Incineration, therefore, is most acceptable where the developed heat can offset other fuel and energy needs. Adsorption, absorption, and condensation techniques, although effective, are limited to exhaust streams with a much narrower range of process characteristics than those that are required for incineration.

Process and material changes are the most diverse options available to surface coating industries. Among the available process and material changes are:

1. New coating technologies, e.g., waterborne, high solids, and powder coatings
2. Reduction of air ingestion into the gas stream requiring treatment
3. Curing coatings in an inert gas
4. More efficient coating methods

Although these changes offer great promise, each one is unique. Consequently, the number necessary to meet all product and process requirements is large, and conversion costs are frequently high. Process and material changes, therefore, can often be implemented over much longer time periods than those required for installing add-on devices.

Several factors influence the effectiveness, cost, and applicability of available control devices or techniques to a given source category. Often the characteristics of a particular process or exhaust gas stream dictate the use of certain control techniques. Many control methods are equal in reducing pollution, but vary in cost.

Other factors that are unique to the control of organic emissions influence the selection of a control option. For example, most organic compounds are derived from petroleum, so that the increasing cost of crude oil provides considerable economic incentive both to reduce solvent consumption and to maximize recovery.

Insurance and occupational safety requirements that specify maximum organic concentrations for fire prevention and operator safety will dictate or preclude the use of certain options.

Finally, long-term warranties or customer requirements can limit the scope of material or process changes.

COATINGS TO REDUCE VOLATILE ORGANIC COMPOUND EMISSIONS

Methods other than waterborne coatings for reducing organic emissions are high solids and powder coatings (see Tables 1.4 and 1.5), and the recently developed aqueous powder suspensions.

High Solids Coatings

High solids coatings contain 50% or more solids by volume.

One method of obtaining high solids coatings is to use lower molecular weight polymers which require less solvent to attain the desired application viscosity.

TABLE 1.4. REDUCTION OF SOLVENTS IN COATINGS.

	Percent Organic solvent	Remarks
Waterborne coatings	10	Adapted to most equipment
High solids coatings	< 30	Viscosity problems, limited systems
Powder coatings	None	High cost, need new equipment

TABLE 1.5. CONTROLLING VOLATILE ORGANIC COMPOUND EMISSIONS.

Type of control	Reduction of solvent, %
Incineration—Catalytic	90
Noncatalytic	90–98
Carbon adsorption	90
Waterborne coatings	60–90
High solids coatings	60–90
Powder coatings	100
Ultraviolet curing	up to 100

Another method of reducing the viscosity of high solids coatings is by heating the coating material. Usually an increase of temperature from 70°F to 125°F is equivalent to 10% solvent reduction. However, heating can cause loss of solvent crucial to the application performance of high solids coatings. Heating can also cause premature gelling, particularly on standing. Many two-component systems use a catalyst to increase the rate of the curing reaction. Although these chemical reactions can take place at room temperature, many operations use low temperatures to cure two-component systems rapidly so that the coated product can be handled sooner. The oven temperatures required are much lower than for conventional ovens, and the amount of solvent evolved is less. This would result in significant energy savings.

Fast-reacting two-component systems are usually applied with special spray guns that mix the two components at the spray nozzle. Coatings of high viscosity can be applied with a knife coater.

Advantages

1. High solids coatings in most cases can be applied on conventional equipment; therefore, conversion costs are low.
2. In many cases, the energy required for curing is less than for either conventional solvent coatings or waterborne coatings.

Most high solids resins fall into two categories, two-component ambient-temperature-cured and single-component heat-cured. The most important types are:

Two-component—ambient cure	*Single-component—heat cure*
Urethanes	Epoxy
Acrylic urethane	Acrylic
Epoxy-amine	Polyester
	Alkyd

Disadvantages

The limitations of high solids relate to the properties and availability of these coatings.

1. Achieving the desired properties in the finished coating is difficult. Most of the polymerization in high solids coatings occur after application, and controlling the conditions so as to produce the desired properties is more difficult.
2. The availability of high solids coatings is very limited.
3. Pot life of two-component systems is short, leading to application difficulties.
4. Toxicity of isocyanates used for urethanes can be a problem.

Powder Coatings

Powder coatings involve the application of finely divided coatings to a surface followed by a melting of the coating solids into a continuous film. Very little solvent is used (less than 1%), and the process is almost pollution-free.

Advantages

1. Single-coat applications with the fluidized bed technique are possible for thicknesses up to 40 mils.
2. Material utilization can approach 100% if the powder can be collected and reused.
3. Safety aspects of powder coatings offer some advantages, since powders are low in toxicity and nonflammable in storage. However, virtually any organic powder suspended in air can be explosive.
4. Maintenance is generally less because the powder can be vacuumed from any unbaked surface.
5. Exhaust air volume is greatly reduced from that used for solvent-borne spray because application is either automatic or else done in a much smaller area. Spray-booth air can be filtered and returned.
6. Water pollution problems are absent because dry particulate collection is possible.

Disadvantages

1. Color change is a difficult problem with powders.
2. Color matching is more difficult with powder coatings than solvent coatings.
3. Powder coatings materials are discrete particles, each of which must be the same color. Thus, there can be no tinting or blending by the user. Color must be available from the manufacturer.

4. The high temperature required for powder coatings makes them applicable only for metals and some plastics.

5. A typical particle size for sprayed powder coatings is generally greater than 15 micrometers. Since 1 mil is about 25 micrometers it is obvious that the uniform spray coatings are difficult to achieve at coating thickness of less than 2 or 3 mils. Fluidized bed coatings are usually about 200 micrometers in diameter and thus are not applicable for thin coatings.

Aqueous Powder Suspensions

A very recent development is aqueous powder suspensions (APS). These are powdered resins in water slurry form which can be applied with conventional spray guns. APS products are completely solvent-free and can be applied in thinner films than powder coatings.

Considerable work has been done with epoxy systems in aqueous suspensions. The APS system is a dispersion of powdered resins, hardeners, and other additives in water. This suspension contains no solvent or emulsifiers, since it is a solid–liquid system.

The process used to manufacture an APS system closely resembles those used to make a powder coating. A mix consisting of resin, pigment, and other additives is extruded, cooled, milled, and sieved to produce a powder comprising 100-micron particles. Then, going one step beyond powder coatings, this powder is finely ground in an aqueous slurry to particle sizes of 10 microns.

APS systems, while developed for application by standard spray guns, can be applied by airless spraying, dip coating, electrostatic spraying, and reverse roller coating.

The APS products have a solids content of from 20 to 30% so that they require more energy for curing than do powder systems or other waterborne coatings.

Compared to powdered coatings, colors can be shaded more easily and colors switched quickly on the production line. The hazard of powder dust is eliminated. The cost is slightly higher because of the additional cost of the wet grinding operation.

This type of product can be used for nearly all types of industrial products—for example, automotive primers, steel furniture, appliances, and can coatings.

PAINT INDUSTRY STATISTICS

Table 1.6 gives paint production and sales figures for the United States for the years 1967–77.

At the present time waterborne paints are about 55% of trade sales paints, a

TABLE 1.6. PAINT PRODUCTION AND SALES, UNITED STATES.

Year	GNP $ Billion	PRODUCTION, MM GALS Trade paints	Industrial finishes	Total	FACTORY SALES, MM$ Trade paints	Industrial finishes	Total
1967	794	398	393	781	1340	1018	2348
1968	865	424	419	843	1428	1159	2587
1969	931	430	450	880	1474	1304	2777
1970	977	428	399	827	1498	1239	2737
1971	1056	431	443	874	1563	1268	2831
1972	1155	452	475	927	1659	1350	3009
1973	1295	430	475	905	1659	1474	3133
1974	1413	473	455	928	1871	1801	2672
1975	1516	451	439	890	2079	1948	4027
1976	1692	473	455	927	2426	2228	4654
1977	1890	429	417	846	2286	2248	4534

GNP = Gross National Product
MM = millions

TABLE 1.7. ESTIMATE OF U.S. CONSUMPTION OF SYNTHETIC LATEX POLYMERS, 1978, MILLIONS OF DOLLARS.

Industry	Acrylics	Poly-vinyl acetate	Styrene-butadiene	Ethylene vinyl acetate	Others	Total
Paints and coatings	$175	$ 95	$ 1	$ 2	Min.	$273
Paper	20	40	110	Min.	Min.	170
Adhesives	20	100	Min.	16	Min.	136
Carpet backing	—	—	100	4	8	112
Textiles	85	Min.	14	—	Min.	99
Nonwoven binders	30	8	10	6	5	59
Waxes and polishes	6	20	4	—	Min.	30
Putty, caulks, and sealants	7	8	Min.	1	—	16
Leather and hides	15	—	Min.	—	—	15
Printing inks	5	—	10	Min.	—	15
Other	17	19	21	6	7	70
TOTAL	$380	$290	$270	$ 35	$ 20	$995

Min. = minor (< 1 million dollars)

figure expected to grow to about 77% by 1985. In industrial coatings about 14% of the present-day coatings are waterborne, with projections to about 31% of the total by 1985. The trends to waterborne products in related fields such as caulks, adhesives, and inks have been comparable.

Table 1.7 gives an estimate of latex consumption in various industries. In

TABLE 1.8. MAJOR U.S. PRODUCERS OF SYNTHETIC LATEX POLYMERS.

Company	Acrylics	Polyvinyl Acetate	Styrene Butadiene	Ethylene Vinyl acetate	Others
Air Products		X		X	X
AMS Co.		X			
Borden	X	X			
Celanese	X	X			
Dow			X		
DuPont	X	X		X	
Firestone			X		X
General Tire			X		X
Goodrich	X		X		X
Goodyear			X		X
W. R. Grace		X	X		X
Monsanto		X			
National Starch	X	X			X
Reichhold	X	X	X		
Rohm & Haas	X				
Union Carbide	X	X		X	X

the coating industry and textiles, acrylics are number one; in paper coating and carpet backing, styrene-butadiene latices predominate; whereas in adhesives the leader is polyvinyl acetates.

Table 1.8 lists major U.S. producers of synthetic latex polymers.

REFERENCES

The Kline Guide to the Paint Industry, Fairfield, New Jersey: Charles H. Kline & Co.

Martens, Charles R., *The Technology of Paint, Varnishes & Lacquers,* Huntington, New York: Krieger Publishing Co. (1976).

Paint Red Book—1978, New York: Palmerton Publishing Company.

2

Properties of Water

Before proceeding very far with waterborne coatings, one must consider the unusual physical properties of water. It is a chemical compound, H_2O, which exists on the earth in three physical states: solid (ice), liquid (water), and gas (water vapor).

Considering that water has a low molecular weight (18), it has a relatively high boiling point. Methane, CH_4, for example, has a molecular weight of 16, very close to that of water; yet methane exists as a gas at ambient conditions. Also hydrogen sulfide, H_2S, with a molecular weight of 34, is a gas. This marked difference in the properties of water suggests that water has unusual physical characteristics.

Forces that exist between molecules are known as van der Waals forces, which are the forces responsible for the existence of substances in liquid form. In the bulk of a liquid below the surface, the molecules are sufficiently close so that the effect of attractice forces is considerable. The attractive forces are sufficiently great to keep all but a small number of molecules from escaping into the vapor state. The forces of attraction among water molecules, in the bulk of the liquid, are significantly stronger than those of methane. When the temperature of a liquid is raised, the increased kinetic energy imparted to the molecules in the bulk of the liquid will tend to overcome the net attractive force among the molecules. As the temperature of the liquid approaches the critical point, the cohesive forces between molecules approach zero. At this point, surface tension vanishes, and the boiling point is reached. The strong cohesive forces in the bulk of water are responsible for its high surface tension and low vapor pressure. Because of the strong forces between water molecules the actual entities present at room temperature are $(H_2O)_n$, where n is 1, 2, 3, or higher.

Although it has a low molecular weight, water has a depressed vapor pressure so that its evaporation rate is low, relative to other solvents utilized in solvent coating systems. The density of water at 25°C is generally intermediate to that of many organic solvents. Water has a peculiar density behavior with respect to temperature. From 100°C to 4°C the specific gravity of water increases until a maximum is reached at 4°C; from 4°C down to −10°C it decreases. Water expands considerably on solidification: a kilogram of ice at

0°C occupies 1090.8 cubic centimeters, or 90.7 cubic centimeters more than an equal weight of water at the same temperature. Expansion is therefore 9.1%. This is unusual, since nearly all substances contract on changing from a liquid to a solid state.

Pure water is a poor conductor of electricity. Ordinary tap water has a small capacity for conducting electricity because of traces of impurities such as carbon dioxide or metal salts.

Most substances exist in three physical states: solid or crystalline, liquid, and gaseous. Within these states there may be several different forms. For example, sulfur has one vapor state, two liquid states, and several different solid forms. There are no fewer than five different forms of ice, depending on the temperature and rate of freezing. Physical properties of water compared to common solvents are shown on Table 2.1.

SURFACE TENSION

Fluid surfaces, because of these strong attractive forces, exhibit certain features resembling the properties of a stretched elastic membrane. Water, in comparison to other liquids, has a very high surface tension. (See Table 2.2.)

Surface tension is usually expressed in terms of dynes per centimeter. Most emulsion paints, because of the presence of surfactants, have surface tensions in the range of 30 dynes per centimeter. The water-solution types have higher

TABLE 2.1. PHYSICAL PROPERTIES—WATER VS. SOLVENTS.

Property	Water	Mineral spirits	Acetone	Xylene
B.P., °C	100	214.5	56.5	144
F.P., °C	0	−12	−95	−25
(B.P. − F.P.), °C	100	226.5	151.5	169
Solubility parameter	23.5	6.6	10	8.8
H-bonding	39.0	0	9.7	4.5
Dipole moment, Debyes	1.84	0	2.88	0.4
Latent heat of vap., cal/g at B.P.	540	115	135	94
B.P. elevation constant	0.51	2.79	1.33	2.67
Surface tension, dynes/cm	73	18	24	30
Thermal conductivity $\times 10^3$, W/m^2 °C	5.8	1.49	1.8	1.59
Specific heat	1.0	0.52	0.5	0.4
Density, g/ml	1.0	0.751	0.787	0.86
Refractive index Na^d/20 °C	1.3330	1.4216	1.362	1.50
Dielectric constant	78	1.83	21.3	2.37

TABLE 2.2. SURFACE TENSION OF PURE LIQUIDS.

	Dynes/cm
Mercury	485.0
Water	72.8
Acetylene tetrabromide	49.7
Nitrobenzene	43.4
Nitromethane	36.8
Oleic acid	32.5
Carbon disulfide	31.4
Benzene	28.8
Toluene	28.4
n-Octyl alcohol	27.5
Chloroform	27.1
Carbon tetrachloride	26.6
Methylpropylketone	24.1
n-Octane	21.7
n-Hexane	18.4
Ethyl ether	17.1

surface tensions. It is not desirable to add additional surface-active agents to reduce surface tension because the water resistance of the coating may be reduced. Alcohols are sometimes used to reduce the surface tension and then, because of their volatility, leave the coating.

Because of its high surface tension, water has difficulty in wetting some metal surfaces. The "critical" surface tension of a solid, which is defined as being equal to the surface tension of the highest-surface-tension liquid that will wet the solid surface, is a useful way of characterizing a solid in terms of wettability. (See Table 2.3.)

EVAPORATION RATE

Although it has a low molecular weight, water has a depressed vapor pressure, which suggests strongly that its evaporation rate will be slow relative to

TABLE 2.3. CRITICAL SURFACE TENSIONS.

Surface	Dynes/cm
Polytetrafluoroethylene ("Teflon")	18
Polyethylene	31
Tinplate	35
Steel, bonderized	40–45
Alkyd primer	70
Glass (chromic cleaned)	> 70

many solvents utilized in solvent coating systems. See Table 2.4, which lists molecular weight, boiling points, evaporation rates, and heats of vaporization.

Because water (HOH) is a hydroxyl or alcohol-type compound, we might expect it, in comparison to other alcohols, to have a boiling point of about 150°F. However, because of its extremely high polarity, it has a considerably higher boiling point, 212°F. Nevertheless, we would still class it as a low-boiling-point solvent along with ethyl alcohol, ethyl acetate, toluene, VM&P naphtha, and so on. If we compare evaporation rates (i.e., the speed at which a liquid vaporizes at ambient temperatures using ethyl ether as a standard of 1, we find that water is much slower evaporating than other organic solvents in this boiling point range. For example, toluene, which has a slightly higher boiling point than water, will evaporate in about one tenth the time of water.

In baking-type finishes the heat of vaporization (the heat required to vaporize the solvent) is a significant factor. It depends on the polarity of the compound. The heat of vaporization of water, which is 540 calories, is about 2½ times that of ethyl alcohol, 5½ times that of ethyl acetate, and 7 times that of toluene or xylene.

TABLE 2.4. PHYSICAL CONSTANTS OF SOLVENTS AND DILUENTS.

	M.W.	B.P., °F	Evap. Rate	Heat of vaporation, cal/g
Low boiling point				
Ethyl ether	74	94	1	—
Ethyl acetate	88	171	2.7	102
Ethyl alcohol	46	173	7	204
Methyl ethyl ketone	72	175	2.7	—
Water	18	212	30–40	540
Toluene	92	231	4.5	87
Butanol	74	245	20	141
Cellosolve[a]	90	275	28	—
VM&P naphtha	—	240–325	7	—
Medium boiling point				
Xylene	106	286	9	82
Cyclohexanone	99	313	22	—
Cyclohexanol	100	322	150	108
Diacetone alcohol	116	336	60	—
High boiling point				
Tetralin	132	380	700	—
Benzyl alcohol	108	401	500	112
Isophorone	138	419	200	—
Carbitol acetate[a]	176	424	830	—

[a]Registered trademark, Union Carbide Corp.

These unusual characteristics of water must be taken into account especially in handling waterborne baking finishes. A greater time should be allowed between the application of paint and the state of baking. Evaporation can be accelerated by passing air across the surface at temperatures slightly above room temperature. Because water is a relatively low-boiling-point solvent, any water trapped in the coating can cause blistering when baked at temperatures above 212°F. The oven will require a higher heat capacity because of the higher heat of vaporization of water. The higher the humidity, the lower the evaporation.

Figure 2.1 shows the effect of humidity on the evaporation of solvents. Note that an increase in humidity does not affect the evaporation of xylene, whereas with water the evaporation rate is 20 times slower at 90% relative humidity than at 10% relative humidity.

Because the vapor pressure of water increases rapidy with increasing temperature, the atmosphere holds more moisture at higher temperatures (Figure 2.2). At 25°C air saturated with water vapor contains 25 g water/m^3 air. At 60°C the capacity of air for water vapor increases fivefold to 130 g/m^3. At the boiling point of water, air saturated with water vapor holds 600 g/m^3 or 24 times the capacity at 25°C.

Relative humidity (R.H.) is the percentage of water vapor present in air at a specified temperature relative to the saturation concentration at the same

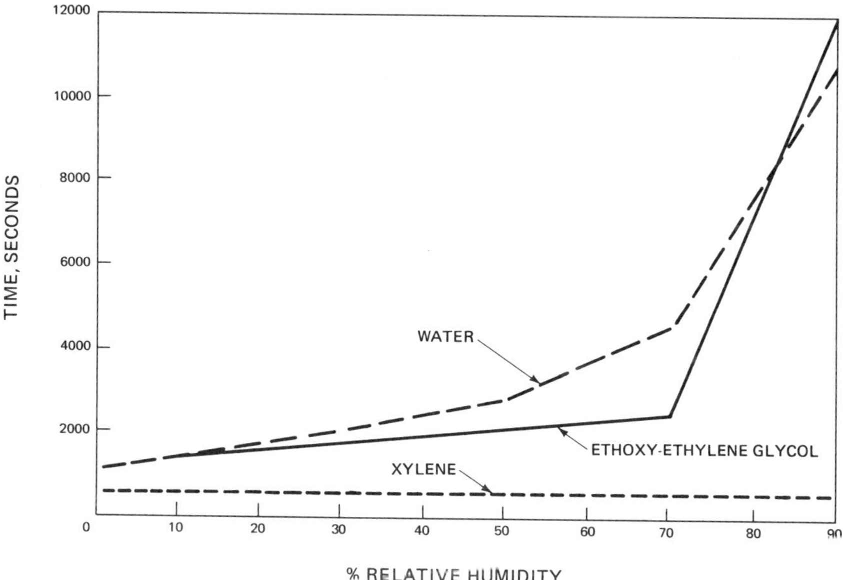

FIGURE 2.1. Effect of humidity on evaporation of solvents. Temperature 25°C; 90% evaporation.

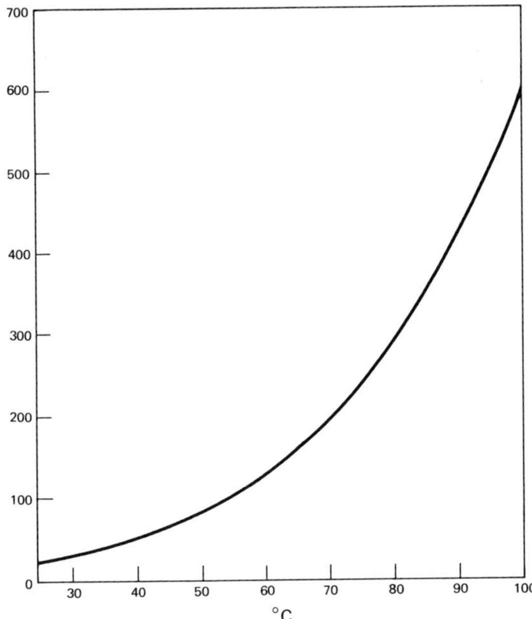

FIGURE 2.2. The capacity of air for water vapor (g water vapor/m^3 air) at saturation vs. temperature.

temperature. Because of the increasing capacity of air for water vapor with increasing temperature, as shown in Figure 2.3, the relative humidity of air containing a fixed amount of water vapor decreases rapidly with increasing temperature. Air saturated with water vapor (i.e., at 100% R.H.) at 25°C is only half saturated (i.e., at 50% R.H.) at 25°C. The relative humidity of the same air at 60°C is less than 20%.

The evaporation rate (E.R) of water is governed by temperature, the relative humidity, and the flow rate of the air across the water's surface. At a fixed temperature and flow rate of air above the water's surface, the evaporation of water is a linear function of relative humidity varying from zero at 100% R.H. to a maximum at 0% R.H.

$$\text{E.R.} = C\left(1 - \frac{\text{R.H.}}{100}\right)$$

As a result of the increasing capacity for water vapor, the evaporation rate of water increases sharply with increasing temperature.

Higher air speeds significantly increase the evaporation of water at low and medium relative humidity; yet as the air reaches saturation, the evaporation of water approaches zero no matter what the air speed.

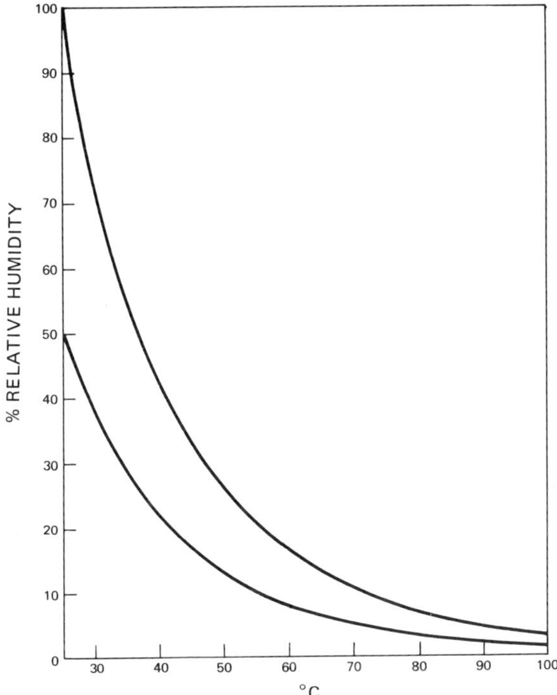

FIGURE 2.3. The change in relative humidity of air, initially at 50% and 100% R.H., with increasing temperature.

SOLVENT ACTION

Water has unique solvent capability that is unmatched by other organic solvents. It is able to form relatively concentrated solutions of inorganic compounds such as sodium chloride, copper sulfate, ferric chloride, and the like. Many organic compounds can also be dissolved in water. Organic solvents, on the other hand, lack the capacity to dissolve significant amounts of inorganic compounds.

Water has the ability to dissolve gases, and with carbon dioxide, forms carbonic acid, H_2CO_3. This will lower the pH of the water.

The ability of water to dissolve inorganic matter accounts for the fact that water from lakes and wells can contain various amounts of minerals

Hardness of water is usually due to trace solubility of magnesium and calcium salts in the form of sulfates, carbonates, and chlorides. Other metals such as iron may also be found in water.

The matter of water hardness is important in waterborne finishes, particularly when water is added to the paint prior to application. Extemely hard water with pH extremes causes serious deviations in application, pot life, and dry film performance.

pH

pH is the unit of measure used to express the degree of acidity of a substance. The pH scale goes from 0 to 14. A pH of 0 means a very high acid activity, whereas a pH of 14 means a very low acid activity or high alkalinity. The midpoint between these extremes is a pH of 7, which is neutral. This is the pH of pure water.

pH is the logarithm of the reciprocal of the hydrogen ion concentration. A unit change in pH corresponds to a tenfold change in acidity. Table 2.5 gives the pH of some common materials.

To understand more about pH we need to know more about the chemistry of water. A molecule of water is composed of one oxygen atom and two hydrogen atoms and looks something like this:

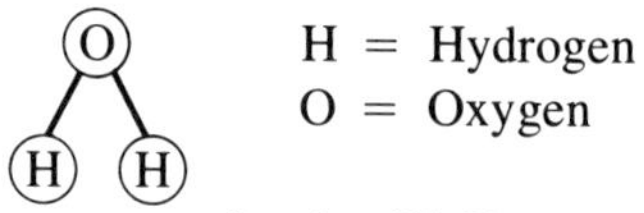

H = Hydrogen
O = Oxygen

water molecule (H_2O)

In pure water, most of the water molecules remain intact. However, a very small amount of them react with each other in the following manner:

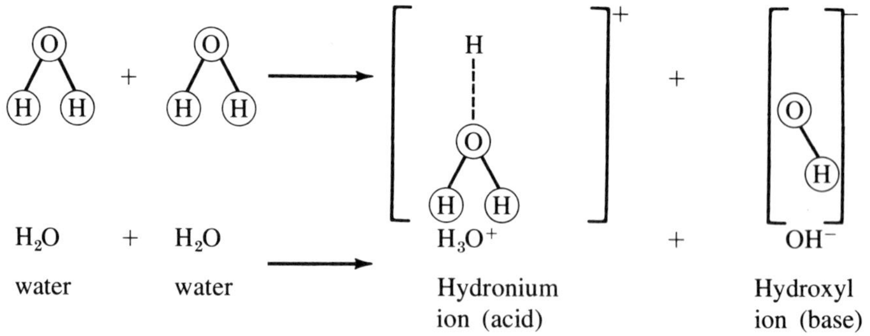

$$H_2O + H_2O \longrightarrow [H_3O]^+ + [OH]^-$$

H_2O	+	H_2O	$\longrightarrow$	H_3O^+	+	OH^-
water		water		Hydronium ion (acid)		Hydroxyl ion (base)

The hydronium ion (H_3O^+) is the chemical unit that accounts for the acidic properties of a solution. The hydroxyl ion is the chemical unit that accounts for the basic or alkaline properties of a solution. When pure water reacts, it produces an equal amount of H_3O^+ and OH^-. Thus it does not have an excess of either ion, and it is called a neutral solution.

If a strong acid such as hydrochloric acid (HCl) is added to water, it reacts with some of the water molecules as follows:

TABLE 2.5. pH OF SOME COMMON MATERIALS.

pH	Material
< 1	5% Sulfuric acid
2.3	Lemon juice
3.5	Orange juice
6.5	Milk
7.0	Water
8.0	Sea water
9.2	Borax
10.5	Milk of magnesia
13.0	Lime ($Ca(OH)_2$)
14.0	Lye (4% NaOH

$$HCl + H_2O \longleftrightarrow H_3O^+ + Cl^-$$

Thus the addition of HCl to water increases the H_3O^+ or acid concentration of the resulting solution.

If a strong base such as sodium hydroxide (NaOH) is added to water, it ionizes as follows:

$$NaOH \longleftrightarrow Na^+ + OH^-$$

Thus the addition of NaOH to water increase the OH^- or alkali concentration of the resulting solution.

Another interesting aspect of water chemistry is that the concentrations of H_3O^+ and OH^- remain in balance with each other. An increase in the concentration of H_3O^+ causes a proportional decrease in the concentration of OH^-. This relationship is shown in Table 2.6. Note that:

1. As the (H_3O^+) concentration decreases, the pH increases.
2. As the acid (H_3O^+) concentration decreases, the base (OH^-) concentration increases proportionally.
3. At pH 7 the acid (H_3O^+) and the base (OH^-) concentrations are equal. This is the neutral point.
4. The pH scale represents the number of places the decimal point is moved to the left of one in expressing the acid (H_3O^+) concentration.
5. Each pH unit represents a tenfold change in H_3O^+ or OH^- concentration. For example, a solution at pH 6 is ten times more concentrated in H_3O^+ ions than a solution at pH 7.

The pH of a waterborne paint has a profound effect on the physical properties and performance of the system. Each latex system is more stable in a specific pH range. Surface-active agents have pH ranges in which they are

TABLE 2.6. ION ACTIVITY.

pH	(Mole/liter) H_30^+ (acid)	OH^- (base)
0	1.0	0.00000000000001
1	0.1	0.0000000000001
|	|	|
|	|	|
|	|	|
|	|	|
7	0.0000001	0.0000001
|	|	|
|	|	|
|	|	|
|	|	|
13	0.0000000000001	0.1
14	0.00000000000001	1.0

more effective. Many protective colloids such as carboxy methyl cellulose, casein, poly acrylates, and alginates are soluble only at pH above 7 and become insoluble at a lower pH. A very high pH will hydrolyze materials such as protein, alkyds, and so on. Maintenance of a high pH in the package keeps in check certain bacteria which might cause spoilage. pH also affects the solubility of water-reducible resins. The use of permanent alkalis such as sodium hydroxide should be avoided if possible, since water-soluble materials are left in the film, thereby hurting water resistance. If possible, volatile alkali such as ammonia should be used for adjusting the pH of the paint. Since ammonia is volatile, it can be lost from the batch during manufacture and cause difficulties such as thickening and finally coagulation of the paint.

NONCOMBUSTIBILITY OF WATERBORNE COATINGS

Since they contain large amounts of water, waterborne paints are classified as noncombustible by most standards. Because they contain no solvents, latex paints will neither flash nor burn. However, watersoluble coatings in which the organic solvent may be 20% of the solvent mixture will have a closed-up flash point similar to that of the organic solvent, but will not support combustion.

Most safety regulations concerning the storage of flammable liquids are based largely on the flash point of the material. Many waterborne coatings have flash points comparable to those of solvent-borne coatings. The flash point of waterborne paint is usually close to the flash point of the most volatile solvent. Since the prodominant volatile component of waterborne

TABLE 2.7. OPEN-TANK FLAMMABILITY.

Paint	Closed-cup flash point, °F	Comments
80% water, 20% solvent	90	Paint did not ignite or burn.
65% water, 35% solvent	140	Paint did not ignite or burn.
100% solvent	82	Paint ignited and burned.

paints is water, the closed-cup flash point does not give an accurate indicator of the fire hazard.

Ignition of a flammable liquid is dependent upon obtaining a concentration of flammable vapor in the air over the liquid surface that exceeds the lower flammability limit (LFL). In open tanks waterborne paints containing 20 to 35% organic solvent do not ignite or burn (see Table 2.7). The possible explanation is that waterborne coatings have a lower rate of vaporization of flammable solvent than solvent-borne coatings. Natural air circulation across the liquid surface may prevent the formation of a flammable vapor–air mixture for waterborne coatings. Water vapor may make inert the vapor space above the liquid surface. Also, with the large amount of water present, heat is absorbed and dissipated. Tests show that even though the waterborne paint has a closed-cup flash point, it will not ignite or burn in an open tank.

REFERENCES

Evaluation of the Fire Hazard of Water-borne Coatings, Factory Mutual Research—National Paint & Coatings Association, Scientific Circular 804.

3

General Properties of Coatings

General properties of coatings are discussed in this chapter. The type of binder determines the flexibility, chemical resistance, and durability of the coatings. The pigment determines the color and color stability, while the amount of pigment determines the gloss and permeability of the coating. The solvent or dispersing medium determines the application properties such as flow and to some extent the drying time.

PACKAGE PROPERTIES OF PAINTS

Package properties are concerned with how the paint appears and the stability in the package.

Viscosity

The measurement of viscosity or rheology of a paint is important, as the viscosity affects the application properties of coatings.

Newtonian flow is of the type occurring in a liquid system where the role of shear is directly proportional to the shearing force:

$$N = \frac{f/A}{dv/dx} = \frac{\text{shear stress}}{\text{shear rate}}$$

Low solids paints used for dipping and spray application are closest to this type of viscosity. Efflux viscosity devices such as Ford or Zahn are used to measure this type.

Non-Newtonian flow is plastic, pseudoplastic, or dilatant flow. This is the type of flow encountered in paints used for brushing applications. Here measuring devices such as the Krebs-Stormer are used.

Skinning

Oxidizing-type solvent paints tend to form an insoluble skin on the surface that is exposed to the air. This usually does not happen with waterborne paints.

Settling

Settling is the degree to which the pigment settles in the container over a period of time. The observation is generally rated on a 10–0 scale with 10 being no settling. Settling is classified from soft to hard settling, depending on how difficult it is to redisperse the pigment.

Weight per Gallon

This measurement is made to determine the conformity of the paint to the formula and is therefore a measure of quality control. The measurement is made by a weight per gallon cup.

Flash Point

Flash point is the measurement of the lowest temperature at which the vapor will ignite when brought into contact with spark or flame. The flash point is required information for meeting shipping, storage, and use regulations. It is generally measured by the Tag open cup, the Tag closed cup, or the Pensky-Martens closed cup tester (for pigmented materials).

Freeze–Thaw Stability

This test is performed on waterborne paints only and is used to determine the package stability of the product when subjected to cycles of freezing and thawing during storage and shipment.

Fineness of Grind

This is a measurement of the degree of dispersion of the pigment and is most frequently measured by the Hegman North Standard gauge.

APPLICATION PROPERTIES OF PAINTS

These properties involve ease of brushing, rolling, or spraying. Some specific properties and tests are:

Leveling

This is a subjective determination of the manner in which the paint flows out to form a uniform, smooth level surface. It is generally rated on a scale of 10–0, with 10 being excellent leveling.

Sag

The property measured is the ability of the paint to resist sags or curtains upon application to a vertical surface. Sag testers such as the Lenata or Baker are used for this measurement.

Spatter

This property is measured on paints that are applied by roller. A black plastic panel is placed on the roller handle directly behind the roller to catch the spattered paint during a controlled application. The degree of spatter is rated from 10 (no spatter) to 0 (very severe spatter).

Drying Time

This is the determination of the speed at which a paint dries. Set-to-touch (a measure of the paint's wet edge time), tack-free time, dry-hard, and dry-through are the properties measured.

Film Formation

Film formation from the wet paint to the dry film can be accomplished in several ways. In some cases, a combination of two or more of these methods is involved.

Solvent Loss

The solvent evaporates from the surface, and the paint becomes solid. The simplest form of this would be a lacquer which dries by the evaporation of solvent.

Air Reaction

The binder of the coating reacts with the oxygen of the air to form a cross-linked film. Drying oils such as linseed would be of this type. Drying oils are incorporated into alkyds, epoxies, urethanes, and so on, to accomplish this.

Coalescence

Emulsion paints undergo coalescence on drying. As the water leaves the coating, the droplets of binder and pigment come closer together and finally fuse into a continuous film.

Chemical Reaction

In this case the coating contains two or more reactive materials which will react to form a cross-linked film. An example of this are epoxies and amines which are mixed prior to application and will cure at room temperature. Alkyd–amino systems are stable at room temperature, but cure upon baking.

Fusion

Powder coatings are examples of coatings that cure by fusion. The powder coatings are applied as a powder to the surface and then melted into a continuous film. Water suspension coatings of hard polymers are sometimes applied to metal surfaces and fused by heat.

Film Formation in Waterborne Coatings

In solution coatings individual polymer or resin molecules are surrounded and plasticized by water and the other solvents present. The polymer molecules are separated from neighboring molecules by molecular dimensions. With polymers of high molecular weight the molecules are thoroughly entangled. Upon evaporation of solvent the flexibility, mobility, and size of the molecules allow the shrinking system to compact and form a continuous uniform film.

Film formation from a dispersion or an emulsion is more complex. Polymer dispersions consist of a separate polymer phase in the form of individual spheres dispersed in a liquid medium. Such a system is not homogeneous on a molecular basis. As the water evaporates or is absorbed into a porous substrate, the spherical particles come closer and closer together and finally touch. Then in order to obtain a continuous polymer film, free of voids, deformation of the spheres is necessary. This requires the existence of a driving force of sufficient magnitude to overcome resistance of the polymer spheres to this change in shape. Hard polymers will resist deformation very strongly under imposed stresses, whereas rubbery polymers will deform more readily. The potentiality for film formation of a dispersion polymer is related to this deformability. If the polymer is so rigid that deformation does not occur, film formation is not accomplished, and a powdery or spongy structure remains after water evaporation.

With polymer dispersions the capillary pressure of the water exerts forces that pull the dispersed particles together. This pressure is similar to the pressure that is present when two plates of glass are separated by a thin layer of water. Comparable forces exist when two or more spheres are wet by the intervening aqueous phase. As the water evaporates and the spheres come closer together, this pressure increases. When the spheres touch, the evaporation of the water exerts pressure, deforming the spheres, which then form a continuous film. Figure 3.1 shows film formation from a dispersion.

Factors Affecting Film Formation of Waterborne Coatings

There are many factors affecting film formation, which may be environmental, physical, or compositional.

Time

An adequate amount of time is required to allow the film to form properly. Control of the evaporation of water will control the coalescence of the emulsion film. The relative humidity affects the evaporation rate of the water, and the porosity of the substrate also affects the rapidity with which the water leaves the film.

Temperature

Temperature of application is important, and temperature affects the hardness of the polymer, which, in turn, affects fusibility. Obviously the temperature cannot be below the freezing point of the water phase.

Physical Influence

The physical state of the emulsion has an important influence on film formation. Normally, the finer the particle size, the easier it is to obtain good film

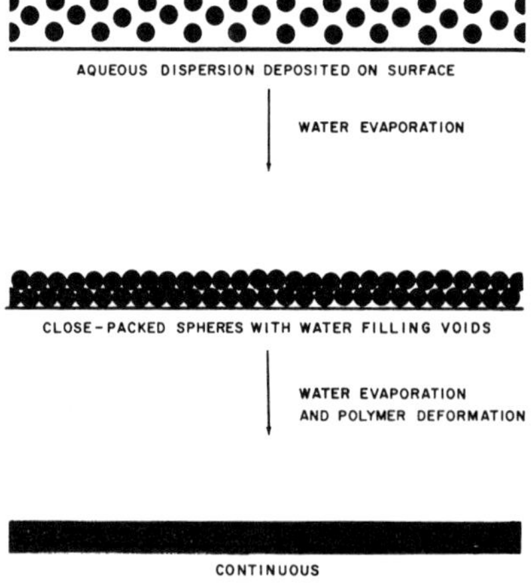

FIGURE 3.1. Film formation from a dispersion.

formation. An emulsion in a partially flocculated condition may trap water, leaving voids in the film.

Composition

The chemical composition and molecular weight of the polymer affect the fusion temperature. If the polymer is too hard to fuse at application temperature, plasticizers or solvents may be added to soften the resin and aid fusion.

Mud-cracking of Latex Paint Films

"Mud-cracking" as a general term applies to intersecting cracks that penetrate at least one coat in a dried paint film. Mud-cracking occurs when the coating becomes rigid before all of the volatiles have left the coating. The phenomena of mud-cracking has been investigated in simple latex paint systems in an attempt to determine the factors, particularly formulation variables, that influence this type of latex film. It has been found that:

1. The lower the temperature, the thicker the film, and the more porous the substrate, the greater the tendency is to mud-cracking.
2. The higher the coating level of TiO_2, and the higher the TiO_2 content, the greater the tendency is to mud-cracking.
3. The higher the PVC, the more susceptible is the coating to mud-cracking.
4. The incorporation of a small portion of platelike extenders, e.g., mica, considerably reduces the tendency to mud-cracking.
5. Increasing the content of an effective coalescent in a latex paint decreases the tendency to mud-cracking.
6. Polymer type or composition does not exert a marked influence on mud-cracking.
7. Minor constituents of the formulation, e.g., thickener, dispersing agent, did not influence the degree of mud-cracking.

DRY PAINT PROPERTIES

Gloss

A high-gloss surface is a smooth surface that reflects most of the light that falls upon it. Rough surfaces or flat surfaces scatter the light. Highly pigmented paints often have low gloss because pigment particles protrude through the surface causing roughness. Gloss is usually measured at an angle of 60° from the horizontal.

Hiding Power or Opacity

The ability of a paint to cover or obliterate the substrate is called hiding power or opacity. Pigments contribute to hiding power by reflecting, refracting, and absorbing the light that enters the paint film. The hiding power of a white pigment is determined by the difference between its index of refraction and the index of refraction of the surrounding medium or binder. This property is generally measured by dividing the reflectance of the paint over a black surface by the reflectance over a white surface and multiplying by 100 to obtain a percentage figure.

Color

The color of the pigment is due to the pigment interacting with light. Ordinary white light is composed of all of the colors visible to the human eye. Pigments have the ability to absorb some of these colors and reflect and transmit others. The color will be the reflected light. A blue paint absorbs all of these colors but blue.

Color is measured on instruments by spectral reflectance and composition on such instruments as the Beckman, Bausch & Lomb, Deano Spectrophotometer, Collmorgen Color Eye, and the Gardner and Hunter Color Difference meters.

FILM RESISTANCE PROPERTIES

The resistance properties of a protective coating are a measurement of its relative ability to withstand attack by any of the conditions to which it is exposed.

Hardness

This property is a measurement of the resistance of a film to indentation or scratching, or of its surface hardness. Indentation is often measured by using the Pfund indentor or Knoop indentor. Scratch hardness can be measured by the Hoffman Scratch Tester, Arko Microknife, and the Beel Scratch-Adhesion Tester. Pencil hardness is also used, with the range being from the softest (6B) to the very hard (9H) lead.

The instrument used most frequently to determine the surface hardness of a coating is a Sward Rocker.

Abrasion

Paint films are frequently subjected to abrasive wear, generally by a mechanical action such as rubbing, scraping, or erosion. Abrasion is measured with a Taber Abraser and the Gardner Falling Sand Abrasion Tester.

Adhesion

Adhesion to the substrate is dependent on both chemical and physical bonding. This is measured by a simple knife test, a razor cross-hatch test, or the Arco Microknife.

Flexibility

A paint film is frequently subjected to movement by contraction and expansion of the substrate due to temperature changes in the case of metals and to moisture changes in the case of wood. Bend tests of painted panels over mandrels are usually used to determine flexibility of coatings.

Impact

In testing impact resistance, a weight is dropped upon a coated specimen to make an indentation in the substrate. Typical instruments used are the Gardner Variable Impacter and the General Electric Impact Flexibility Tool.

Scrubbability

Architectural coatings are expected to be washable and scrubbable. The principal method of evaluating this property uses the Gardner Scrub Tester. (See Figure 3.9.) In the standard scrub test, the paint to be tested is applied on a glass plate at a film thickness of 1.8 mils and allowed to dry for 96 hours. This coating is then tested on a washability machine and should stand 1000 scrubs without wearing through the paint film.

Chemical and Water Resistance

Protective coatings are regularly subjected to environmental attack by water or chemical ingredients which ultimately lead to paint failure. The ability to withstand this attack is measured by immersion tests, condensing humidity, or a blister box.

Durability

The final test for a paint intended for exterior application is its ability to withstand the effects of the weather. Environmental conditions vary from dry rural, to fume-laden industrial, to a marine atmosphere. Durability of a paint film depends on the resistance of a coating to sunlight, moisture, and expansion due to temperature and moisture changes. Ultraviolet radiation degrades the polymer molecules by breaking bonds between atoms. As the binder breaks down, moisture penetrates the coating, accelerating the process. As this proceeds further, enough binder is eroded so that the pigment is exposed and finally freed and lies loosely on the surface. This condition is known as "chalking."

There are various accelerated test procedures indicative of exterior durability. However, the true test of a paint's durability can be measured only by actual exposure to the actual environment on the intended surface.

The Weatherometer is a device intended to simulate exposure to ultraviolet light, rain, and dew.

Salt fog is an accelerated test to determine the relative corrosion resistance of metal protective paints. A 5% solution of sodium chloride is vaporized into a fog at a temperature of 95°F in a closed chamber. The painted panels are exposed for prolonged periods (300 to 1000 hours) and then examined for corrosion.

PIGMENT-VOLUME CONCENTRATION

The type and amount of pigment present affects the performance and properties of the paint. With lesser amounts of pigments, the pigment particles are completely surrounded by the vehicle or binder. As the amount of pigment is increased, a point is reached where there is not enough binder present to fill all voids between the pigment particles. The coating is held together by the pigment's being bonded together at a few spots. At this point, the coating properties undergo a decided change. Properties such as gloss, permeability, tensile strength, and so forth, are affected. This is a volumetric phenomenon, and the relationship can be stated as the pigment-volume concentration (PVC). The pigment-volume concentration is a percentage and represents the pigment volume divided by the total coating solids volume (pigment volume + vehicle solids volume):

$$\text{PVC} = \frac{\text{Pigment volume}}{\text{Pigment volume} + \text{Vehicle solids volume}} \times 100$$

The point at which all voids between pigment particles are just filled with binder is known as the critical pigment-volume concentration (CPVC). M. Van Loo and H. J. Asbeck[4] developed a method by which the CPVC of a solvent paint can be determined by the use of a special filter cell. The pigment is separated from a known volume of paint and the vehicle washed out with solvent. The volume of the voids between pigment is determined by measuring the volume of water required to fill the voids. Figure 3.2 shows the changes in paint properties that occur at the critical PVC.

The CPVC of emulsion paints cannot be determined by direct methods, as the binder cannot be completely separated from the pigment. For this reason

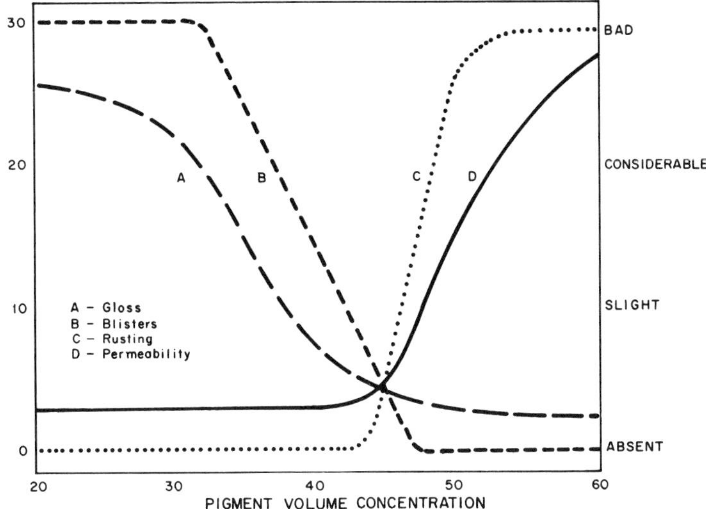

FIGURE 3.2. Effect of CVPC on paint characteristics.

the CPVC of emulsion paints must be determined by indirect methods.[2] Using these methods, it is necessary to make a series of paint samples varying the PVC. Then a series of physical tests is run to determine at what point the physical properties change abruptly.

Becker and Howell[1] presented a method of determining CPVC on emulsion paints by measuring the tensile strength of films. Gloss and permeability of emulsion films were considered unsatisfactory, since the gloss of emulsion paints is usually low and permeability is usually relatively high. (See Figure 3.3.) Other methods of determining the CPVC of emulsion films have been used such as scrubbability and enamel holdout.

See Figures 3.4 and 3.5 for the effect of CPVC on scrubbability and enamel holdout. In this method a porous surface is coated with the emulsion paint the CPVC is exceeded. Liberti[3] determines the CPVC of a vinyl emulsion paint system by a pigment-water sorption method.

Particle Size of Latex. The particle size of the latex has a decided effect on the performance of the latex. As the particle size decreases, the critical PVC goes up; in other words, a finer-particle-size latex has greater pigment binding capabilities. (See Figure 3.6.) Other properties affected by particle size of the latex are shown in Figures 3.7 and 3.8.

PERMEABILITY

Aqueous coatings can be formulated with various degrees of permeability. For this reason emulsion paints are said to "breathe." By this we mean that the

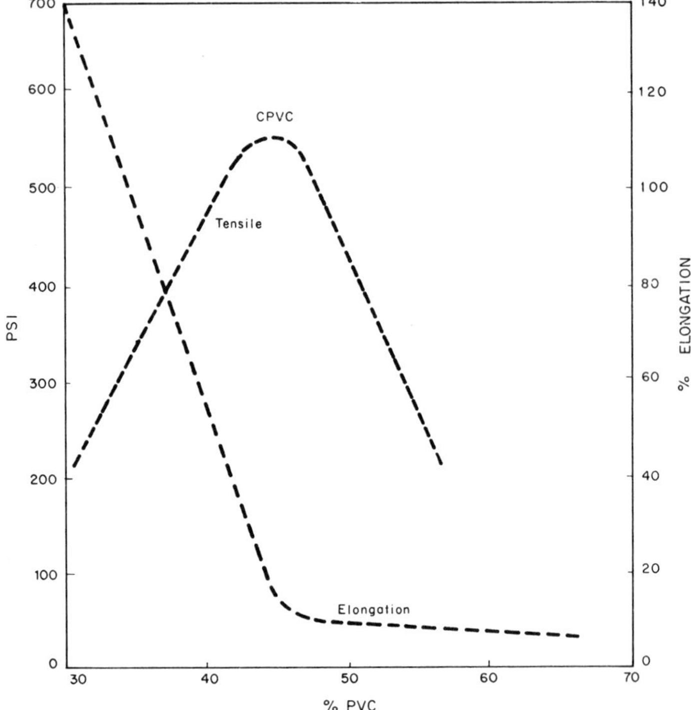

FIGURE 3.3. Tensile and elongation of pigmented emulsion films.

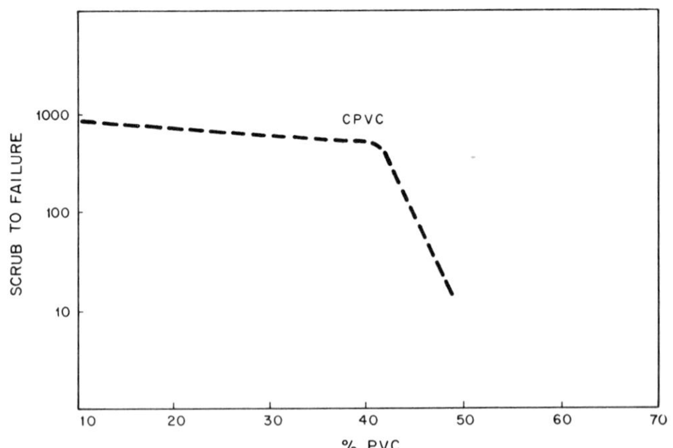

FIGURE 3.4. Scrubbability vs. PVC.

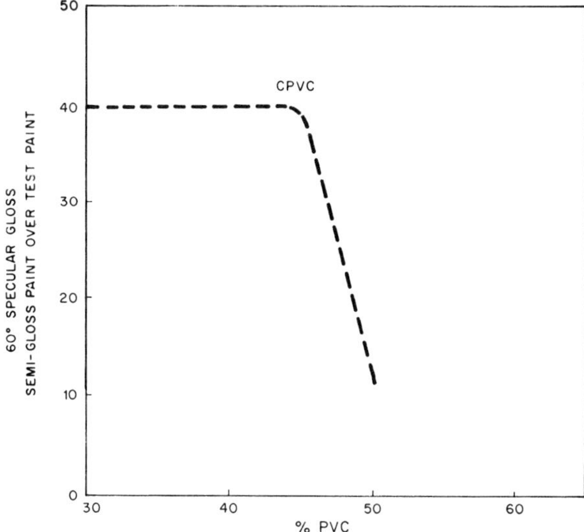

FIGURE 3.5. Enamel holdout vs PVC.

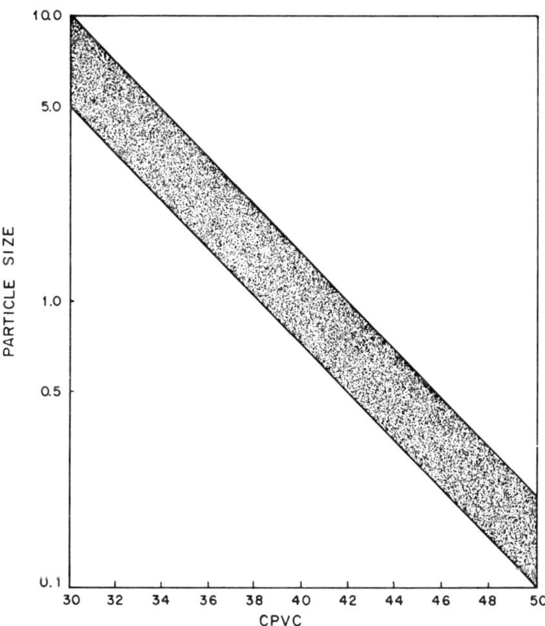

FIGURE 3.6. CPVC vs emulsion particle size.

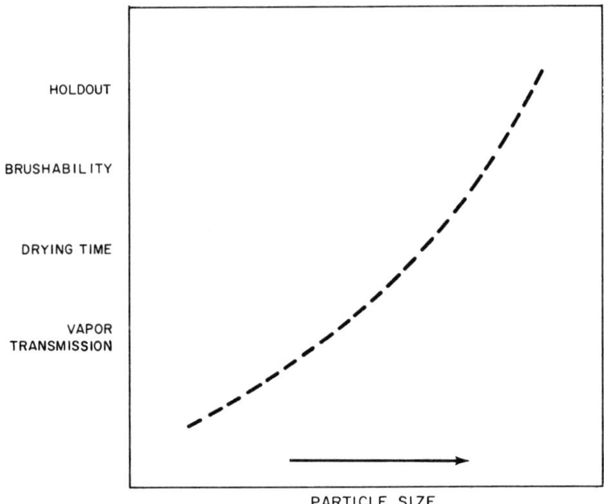

FIGURE 3.7. Properties improved by increased particle size.

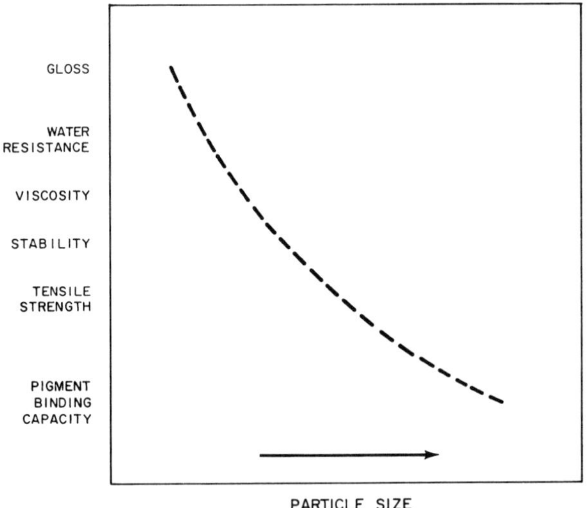

FIGURE 3.8. Properties reduced by increased particle size.

coating allows moisture vapor to pass through. Under certain adverse conditions any paint will blister or peel. Emulsion paints, however, are far less susceptible to blistering and peeling than most solvent paints. Paint peels when moisture vapor condenses and builds up on the substrate, forcing the film away from the surface. Because latex allows vapor to pass through, there is less chance of moisture build-up that will cause the paint to become detached from the substrate.

Conversely, because emulsion coatings are more permeable, it is more difficult to make water-type coatings for metals that will pass salt spray and humidity tests. When a coating applied to masonry surfaces is too porous, efflorescence may occur. This is caused by the migration of soluble salts through the film to leave a white deposit on the surface of the coating.

PENETRATION AND HOLDOUT

Emulsion paints, because of their nature, show less penetration on porous surfaces or, conversely, show good holdout. The higher surface tension of the

FIGURE 3.9. Washability machine. (*Courtesy Gardner Laboratories, Inc.*)

FIGURE 3.10. Latex particles in final stage of film formation. (*Courtesy Dow Chemical Co.*)

TABLE 3.1. POLYMER DISPERSIONS.

Particle diameter, microns	Number of particles per cm^3	Specific surface area of particles, sq cm/g polymer
1	2×10^{12}	6×10^4
0.1	2×10^{15}	6×10^5
0.01	2×10^{18}	6×10^6
0.001	2×10^{21}	6×10^7

water system reduces penetration, since in an o/w emulsion the surfaces are first wet with water which reduces the penetration of the oil phase. The oil phase has relatively large particle sizes and high viscosity, which also contribute to good holdout and bridging.

Dispersion coatings show good holdout on porous surfaces such as paper, textiles, fiber board, wood, fiberglass, asbestos board, concrete block, and so on. The coating is concentrated on the surface. A large-particle-size, high-surface-tension latex will give maximum holdout.

This lack of penetration can be a disadvantage in certain applications. For example, if emulsion paints are used over old chalky paint surfaces, such as old powered paint on masonry surfaces or chalking house paint, the lack of penetration of the emulsion paint may lead to premature failure of the paint system at the chalky interface. If a solvent-type surfacer or primer is used, this will penetrate the chalk and bind it together to form a satisfactory stratum to which paint can be applied.

REFERENCES

1. Becker, J. C., and Howell, D. D., *Offic. Dig. Federation Paint Varnish Prod. Clubs,* **28,** 775 (1956).
2. Conrad, H. A., *Paint, Oil, Chem. Rev.,* Vol. 117, 50–52 (1954).
3. Liberti, F. P., *Offic. Dig. Federation Paint Varnish Prod. Clubs,* **31,** 252 (1959).
4. Van Loo, M., and Asbeck, W. K., *Ind. Eng. Chem.,* **41,** 1470 (1949).

4

Types of Waterborne Coatings

Waterborne coatings can be divided into two general classes, water-soluble or water-dispersible (colloidal dispersions and emulsions). The characteristic that can be used to classify a coating as a solution, dispersion, or emulsion is particle size of the binder.

Definitions of solutions, colloids, and suspensions are somewhat arbitrary, but are usually considered as follows:

	Particle size
Solutions	< 0.001 micron in diameter
Colloids	0.001 to 0.1 micron in diameter
Suspensions	> 0.1 micron in diameter

A solution is a homogeneous mixture of small particles dissolved in a mixture of water and a coupling agent. A water-soluble resin normally contains ionizable amines or carboxylic acid groups that solubilize the molecule. These systems are more easily mixed and applied than other waterborne systems. However, resin properties that make a resin soluble can cause water sensitivity after curing unless additions are made to eliminate this sensitivity. The definition of a solution is "a homogeneous mixture of two or more substances with subdivision to molecular magnitudes."

COLLOIDS

With colloids we are concerned with two-phase systems and with the interface between these phases. With the three states of matter—solid, liquid, and gas—the following interfaces may exist:

Liquid–gas
Liquid–liquid
Liquid–solid
Solid–gas
Solid–solid

Types of colloids are as follows:

Dispersed phase	*Continuous phase*	*Example*
Solid	Gas	Dust
Solid	Liquid	Suspension
Solid	Solid	Ruby glass
Liquid	Gas	Fog
Liquid	Liquid	Emulsion
Liquid	Solid	Gel
Gas	Liquid	Soap foam
Gas	Solid	Foam rubber

Units of measurements for small-size materials and their value in centimeters (cm) are:

	millimeter	micron	millimicron	angstrom
	mm	μ	$m\mu$	Å
Centimeters	10	10^4	10^7	10^8

Table 4.1 shows the relative particle sizes of common materials.

Figure 4.1 shows the relative particle sizes of latices, pigments, and extenders used in coatings.

Figure 4.2 is a photograph of latex particles.

Each of the categories—solution, dispersion, and emulsion—can accommodate a broad spectrum of polymer compositions, and many classes of polymers such as acrylics are available in all types.

Latexes (latices) are high-molecular-weight particles suspended in water by some stabilizing dispersing agent. The resins, of which vinyls and acrylics are the most prominent, have very few functional groups and require emulsifying agents to maintain their form. Latex coatings generally have the best water resistance of waterborne coatings.

An emulsion is a two-phase system consisting of two incompletely miscible liquids, the one being dispersed as finite globules in the other. The dispersed or internal phase is the liquid that is broken up into globules. The surrounding liquid is known as the continuous or external phase. A suspension is a two-phase system closely related to an emulsion in which the dispersed phase is a solid. A foam is a two-phase system closely related to an emulsion in which the dispersed phase is a gas. An aerosol is the inverse of a foam, air being the continuous phase and the liquid being the dispersed phase.

The particle size of the droplets in the internal phase of an emulsion ranges from 0.1 to 10 microns. Within any single emulsion there is a wide range in particle sizes. Particle sizes in an emulsion change with aging, depending on

TABLE 4.1. RELATIVE PARTICLE SIZES.

Range of dimension	State	Visibility	Examples	μ
10^{-1} cm = 1 mm	Coarse dispersion	Visible to naked eye	Fine sand	500
10^{-2} cm		Visible to naked eye	Potato starch	50–100
10^{-3} cm	Fine	Visible with microscope	Coarse particles	45
10^{-4} cm = 1 μ	Emulsion		Butterfat particles in homogenized milk	1–2
10^{-5} cm 10^{-6} cm	Colloid dispersions	Ultra-microscope	Lampblack	0.1
10^{-7} cm = 1 mμ	Solution range	Electron microscope below any visibility	Protein molecule	0.006
10^{-8} cm = 1 Å	Atomic range		Hydrogen	0.0001

the stability of the particular emulsion. On aging, the small particles combine into larger ones. As the particle size increases, viscosity increases.

Water dispersion coatings are intermediate in particle size, in the use of functional groups, and in water sensitivity.

The properties and characteristics generally displayed by coatings in these three categories are summarized in Table 4.2. The relationship of viscosity to solids content is shown in Figure 4.3.

FIGURE 4.1. Particle size comparison (average microns).

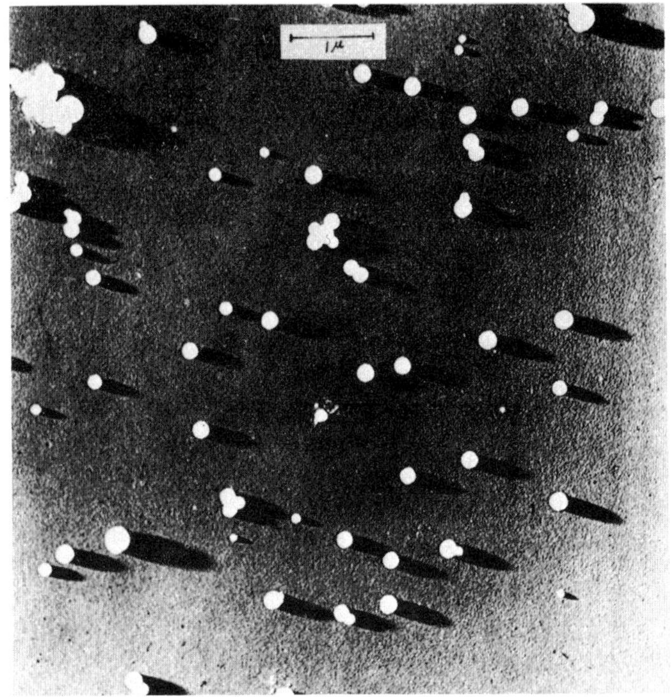

FIGURE 4.2. Magnification of latex particles in colloid dispersion. *Courtesy Dow Chemical Co.)*

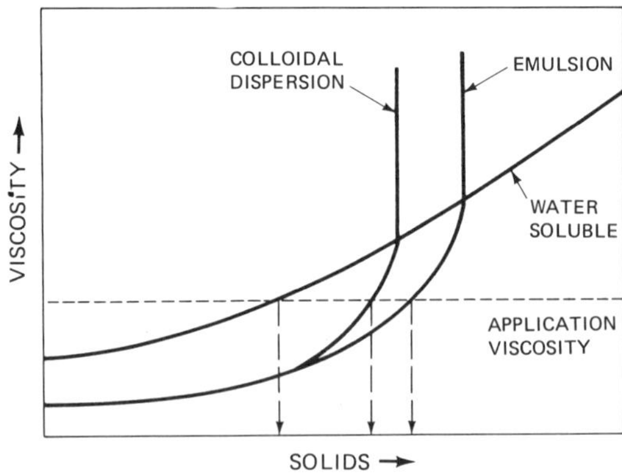

FIGURE 4.3. Relationship of viscosity to solids content in waterborne systems.

EMULSIONS O/W OR W/O

Since we are primarily concerned with the emulsification of oil or hydrophobic materials with water, designations such as oil in water (o/w) emulsions

TABLE 4.2 PROPERTIES AND CHARACTERISTICS OF WATERBORNE SYSTEMS.

	Water-soluble	Colloidal dispersion	Emulsion
Physical			
Appearance	Clear	Translucent	Opaque
Particle size	<0.001 μm	0.01–0.1 μm	0.1–1.0 μm
Molecular weight	5,000–10,000	10,000–50,000	100,000 to 1,000,000
Viscosity	Dependent on molecular weight	Partially dependent on molecular weight	Independent of molecular weight
Formulation			
Pigment dispersibility	Excellent	Fair	Poor
Viscosity control	Dependent on molecular weight	Thickened by addition of cosolvent	Requires thickener
Film formation	Excellent	Good, requires minimum of cosolvent	May require cosolvent
Performance			
Gloss	High	Medium	Low

are used in systems in which the oil is the dispersed phase in the continuous phase water. In water in oil (w/o) emulsions, water is dispersed in oil, which is the external phase.

Whether an emulsion is o/w or w/o can be determined in several ways:

1. The electrical conductivity of an emulsion depends upon the conductivity of the external phase. Therefore, an o/w emulsion will conduct electricity because water is the continuous phase.
2. The dispersibility of an emulsion will determine its type. An o/w emulsion will disperse in water, while a w/o will disperse in oil.
3. Dyes will spread through the external phase. A water-soluble dye will spread through an o/w emulsion, an oil-soluble dye through a w/o emulsion.

MILK

Milk is a colloid or emulsion found in nature. The label analysis of whole milk is:

4%	Fat (in globules)
5%	Carbohydrates (lactose) in clear serum
3%	Protein (casein) in complex micelles
88%	Water
100%	

Milk is remarkably stable as such, and can be concentrated, dried, and reconstituted without alteration of its stable structure.

The colloid structure of milk is worthy of study as a paint model even though the major components have different uses; that is, the fat in milk is for fuel not for polymer structure, and the protein is for structure, whereas in paints protein has poor prospects. Casein micelles have an average diameter of 0.140 millimicrons in a range of about 0.05 to 0.25; this allows a superficial comparison between the blueness of skim milk and monodisperse latexes.

COAGULATION AND GELATION

Emulsion paints, as aqueous colloidal dispersions, must be handled with certain precautions.

They can be coagulated by:

1. Acids
2. Salts, particularly of polyvalent metals
3. Dispersions with oppositely charged particles
4. Water-soluble organic liquids such as acetone or alcohols
5. Freezing
6. Extreme heat
7. Evaporation of water
8. Extreme pressure or shear from mechanical operations such as pumping
9. Extreme dilution
10. Electrical current

Freezing

Paints containing water will freeze when exposed to low temperatures. Because of additives such as glycols, and so forth, which they contain, the freezing point may be somewhat lower than 32°F. When a water-solution-type paint freezes, little damage is done to the system except that the larger volume occupied by the ice may bulge the container. However, in an emulsion system (i.e., an o/w emulsion system), as the external water phase freezes, ice crystals are formed that force together the internal resin phase, causing partial coagulation.

The slower the freezing is, the larger the ice crystals and, therefore, the greater thc damage. The higher the pigment content of an emulsion paint, the less the damage caused by freezing will be. Whereas a low-pigment-content paint may coagulate on freezing, a higher-pigment-content paint might show only a slight increase in viscosity as the result of freezing.

Freeze–thaw cycle tests are made by filling one-pint friction-top cans with paint to be tested and exposing containers to a low temperature of 15°F for 16 hours and a high temperature of 77°F for 8 hours. Paint should pass five temperature cycles without excessive viscosity change.

APPLICATION

Waterborne paints, whether water-soluble or latex, should not be applied at temperatures below 50°F (10°C). This means atmospheric, surface, and product temperatures. Even an air temperature of 50°F (10°C) does not ensure that the walls or surfaces are that warm. The coating will not bond properly to the surface when the temperature is below 50°F (10°C), nor will it form a proper tough film. Product failure may then occur early in the life of the coating. Cracking occurs, and there is a tendency to draw more moisture into the film, owing to improper coalescence or fusion of the film on the surfaces.

An increase in atmospheric humidity slows the drying of waterborne coatings. Movement of air across the surface is important to drying, as the air above the coating becomes saturated with water.

Latex paints are usually faster-drying than water-soluble paints. Most latex products, especially the flats, will dry in about 30 minutes, but higher-gloss products may require two to three hours. Water-soluble products dry on the surface in about 30 minutes, but the final curing may take eight hours or overnight. A painter should not repaint too soon using water-soluble coating systems, but can recoat rather quickly with latex paints.

WATER-SOLUBLE COATINGS

In water-soluble coatings the vehicle or binder is dissolved in water. Upon evaporation of the water, the binder will remain water-sensitive unless converted to an insoluble form by some means such as polymerization. This can be accomplished by incorporating a drying oil in the molecule or cross-linking with an amino resin by baking.

There are essentially three types of water-soluble coatings:

1. The anionic, which contain the functional group $-C\begin{smallmatrix}\nearrow O \\ \searrow OH\end{smallmatrix}$ and can be neutralized (solubilized) with ammonium hydroxide, amines, or sodium or other hydroxides.
2. The cationic resins, which contain the amine functional group and can be neutralized (solubilized) with hydrochloric or acetic acid.
3. Nonionic, which contain solubilizing (OH) groups and can be cross-linked with aldehyde groups.

Advantages

1. Simple formulas—pigment, vehicle, and water with catalyst.
2. Ease of manufacture—can be handled like conventional solvent-type paints with the pigment being ground directly into the vehicle.
3. Good film continuity—continuous film similar to conventional solvent systems.
4. High pigment binding and gloss approaching that of conventional solvent systems. High gloss can be obtained with pigment/binder ratios of 1:1 or greater.
5. Good mechanical stability—coatings based on water-soluble systems can be pumped in all types of equipment similarly to conventional solvent paints. There is a minimum tendency toward skinning or clogging of the dip tank or spray equipment after extended use. Paints can be frozen and thawed repeatedly without adverse effects on the liquid paint.

Disadvantages

1. Usually more costly than solvent systems.
2. Limited number of materials—not as many different vehicle systems available as in solvent systems.
3. Package stability less than for comparable solvent systems.

WATER DISPERSIBLES

In a water-dispersible type of coating the vehicle consists of small particles of binder distributed in a water medium. The size of these particles ranges from 0.1 to 1 micron.

Advantages

1. High molecular weight—possibility of using high-molecular-weight polymers with outstanding properties such as color retention, water resistance, chemical resistance, and good mechanical strength.
2. Air drying—superior coatings obtained on air drying.
3. Fast air drying—tack-free as soon as water leaves the coating.
4. Good holdout—on all types of surfaces including porous substrates.

Disadvantages

1. Complicated formulas—because dispersion systems are heterogeneous systems, they are complicated in makeup.
2. Special manufacture—manufacture is more complicated than for conventional coatings.
3. Stability—the mechanical stability of emulsions is only fair. Excess handling or free-thaw will break an emulsion, which then cannot be restored to its original state.
4. Heterogeneity—because an emulsion contains water-soluble thickener and surfactants, their presence in the film makes it partially permeable to water.
5. Gloss—it is more difficult to produce coatings of high gloss from dispersion systems.

The principal ingredients in a water-borne coating are pigment, binder, and water. There are, however, other ingredients to ensure stability in the container, proper application characteristics, and satisfactory film properties. These are materials such as dispersants, inert or extender pigments, preservatives, antirust agents, fungicides, and so on. If the formula is of the emulsion or dispersion type, then additional materials such as protective colloids, coalescing agents, freeze–thaw stabilizers, and so forth, are present.

A formula may contain as many as 15 to 25 ingredients.

WATER-SOLUBLE FORMULA

Ingredients for a water-soluble-type formula are shown in Table 4.3.

In a water-soluble type of coating, the pigment is ground with a dispersant and a portion of the vehicle. Only enough vehicle is used to produce a heavy

TABLE 4.3. WATER-SOLUBLE FORMULATION.

	Percent
Opaque pigment	20.0
Extender pigment	10.0
Pigment dispersant	0.1
Resin	19.7
Amine solubilizer	1.0
Flow additive	.4
Driers	.2
Cosolvent	9.7
Water	38.8
Defoamer	0.1
	100.0

consistency in order to get good shear in the grinding process. After grinding the remaining materials, such as additional vehicles, driers, antirust agent, defoamers, and so on, are added.

WATER-DISPERSIBLE FORMULAS

The formulation and production of emulsion or dispersion-type products is more complicated. By their physical makeup dispersions are sensitive to mechanical stresses, chemical phenomena, and temperature changes. These factors must be taken into account in their formulation and production. For this reason emulsion or dispersion-type paints and certain additional ingredients are shown in Table 4.4.

Because in an emulsion system the latex cannot withstand a great deal of mechanical stress, the pigments are ground with a dispersant and a protective colloid with a minimum amount of water. The dispersant aids in wetting the pigment and keeping the particles separated. The protective colloid forms an envelope around the pigment particles; it protects the system from coagulation in the presence of an electrolyte or from excessive shear strength. As little as 1% of protective colloid will suffice to envelop sticky polymer particles to keep them from sticking together or coalescing.

The latex is then added to the pigment paste, plus other ingredients such as defoamers, preservatives, buffers, fungicide, and so on. A thickener is then added to adjust viscosity. A defoamer eliminates bubbles or foam; a preservative protects the paint against microorganisms that may be present in raw materials or to which the paint is exposed during manufacture or handling; a buffer prevents pH drift; a fungicide protects the final paint from attack by mildew. A coalescing agent is used to get good film fusion, and is in reality a

TABLE 4.4. EMULSION PAINT.

	Percent
Opaque pigment	20.0
Extender pigment	15.0
Pigment dispersant	0.1
Protective colloid	1.2
Latex	40.0
Preservative	0.5
Fungicide (optional)	—
Coalescing agent	2.0
Defoamer	0.1
Thickener	0.5
Water	20.6
	100.0

volatile plasticizer. Emulsion paints are usually not stable to freezing, and materials are added to protect against freeze damage. Each of these ingredients will be covered in detail in subsequent chapters.

As further progress is made, we will see a combination of latex and water-soluble paints used to obtain superior waterborne paints with the desirable properties of both systems.

5

Organic Binders

There is a wide variety of resins used in coatings, as shown in Table 5.1. Many of them can be produced in a water-dispersion form, which can be accomplished by emulsion polymerization in some cases or by postemulsification of the resin. Some of these resins are water-soluble, while others can be made water-soluble by suitable modification. Many other polymers that are water-soluble are described in Chapter 8. Many of the resins in Table 5.1 can be blended together to give a combination of properties in coatings.

SYNTHETIC RESINS

Synthetic resins are polymeric substances of high molecular weight made up of repeating molecular structural units, usually in the form of chains, with a terminating group at each chain end. As many as a hundred thousand or more repeating units may be combined into a single polymer molecule having a molecular weight in the millions. The most useful high polymers, however, are those with average molecular weights ranging from a thousand to several hundred thousand, corresponding to the degree of polymerization.

The polymer chain may be linear, branched, or cross-linked, or some combination of these forms, depending on the functionality and reactivity of the monomers from which they are formed and on the manner of polymerization. The polymer consists of a closely entangled mass of kinked, coiled, and ordered polymer chains. The physical properties of the bulk polymer are largely determined by the structure of the polymer chains, their molecular weight, and the nature of the intermolecular forces (van der Waals forces) that are in effect between chains in close proximity. The chemical properties of the polymer depend upon its chemical composition and structure.

POLYMER STRUCTURE AND PROPERTIES

Linear Polymers

The solid polymer consists of a closely entangled mass of coiled polymer chains.

TABLE 5.1. COATING RESINS.

Water-dispersible	Water-soluble
Acrylics	Acrylics
Alkyds	Alkyds
—	Urea formaldehyde
	Melamine formaldehyde
Bituminous	—
Cellulose nitrate	—
Cellulose acetate	—
Cellulose butyrate	—
—	Methyl cellulose
—	Hydroxy ethyl cellulose
—	Carboxymethyl cellulose
Ethyl cellulose	—
Drying oils	Drying oils
Epoxy	—
Fluorocarbons	—
Hydrocarbon	—
Phenolic	Phenolic
Polyamide	—
Polyester	Polyester
Polyethylene	—
Polyurethane	—
Polyvinyl acetate	—
Styrene-butadiene	—

In linear polymers the molecular chains, although closely entangled, are separate entities. Hence they are relatively free to slip past each other (plastic flow or permanent deformation). Under the influence of applied stresses they acquire increased mobility (soften) as the temperature is raised, and they can be molecularly dispersed (dissolved) in suitable solvents. Linear polymers, therefore, are permanently fusible and soluble and are described as thermoplastic materials.

Polymers may be characterized as crystalline (ordered) or amorphous (disordered). When the polymer chains have a regular molecular configuration that is conducive to alignment and close packing, the polymer exhibits crystalline properties. Close proximity of the polymer chains in crystalline regions gives rise to large intermolecular forces which result in greater stiffness, greater tensile strength, higher softening point temperature, and increased resistance to swelling and solvation.

Crystallinity in a polymer mass may be induced or increased by stretching. Typical crystalline polymers are polyethylene, polypropylene, polyvinyl chloride, polyvinylidene chloride, polytetrafluoroethylene, and polyamides.

Among typical amorphous polymers are polystyrene, acrylic resins, and copolymers in general, in which the random and irregular spacing of the various monomeric species prevents the close packing of polymer chains. Crystalline polymers are of little interest for coating applications.

Branched Polymers

Branched polymers are those in which basically linear molecular chains acquire long-chain side groups as a result of the random activation of sites along the already polymerized main chain or by the introduction of occasional trifunctional groups during the polymerization process. The side chains are similar in molecular structure to the parent chain and may grow to considerable length with side chains of their own, so that a highly branched structure with properties resembling cross-linked polymers may be built up.

Cross-linked Polymers

In cross-linked polymers the molecular chains are tied together, and their mobility is reduced by the many primary valence bonds between chains. Such polymers are called thermoset materials and are infusible and insoluble except by chemical attack or degradation. The degree of stiffness and associated properties exhibited by a cross-linked material will depend on the extent of the cross-linking. The cross-links serve to limit chain slippage, permitting the material to exhibit rubbery or elastic properties. Such a material may swell markedly in certain solvents yet not dissolve in them. Highly cross-linked materials, on the other hand, exhibit a high degree of stiffness, infusibility, and insolubility. In coatings, polymers are cross-linked after the coating is applied.

Copolymers

A product formed by the polymerization of two or more different monomers is called a copolymer. An example of this would be vinyl chloride–acetate.

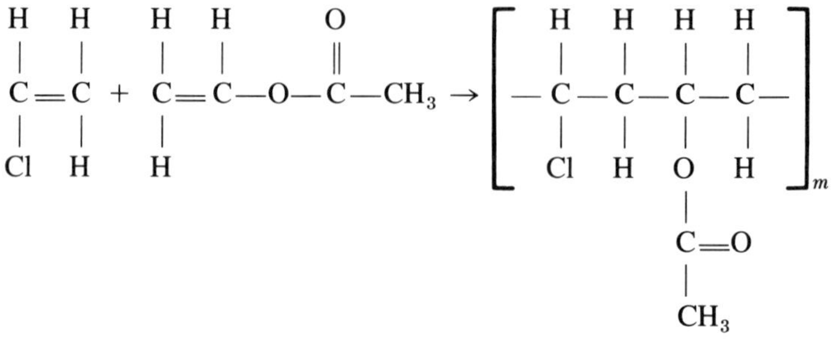

$$
\begin{array}{ccc}
\mathrm{H} & & \mathrm{H} \\
| & & | \\
\mathrm{C} & = & \mathrm{C} \\
| & & | \\
\mathrm{Cl} & & \mathrm{H}
\end{array}
+
\begin{array}{ccccccccc}
\mathrm{H} & & \mathrm{H} & & & & \mathrm{O} & & \\
| & & | & & & & \| & & \\
\mathrm{C} & = & \mathrm{C} & - & \mathrm{O} & - & \mathrm{C} & - & \mathrm{CH_3} \\
| & & & & & & & & \\
\mathrm{H} & & & & & & & &
\end{array}
\rightarrow
\left[\begin{array}{ccccccc}
\mathrm{H} & & \mathrm{H} & & \mathrm{H} & & \mathrm{H} \\
| & & | & & | & & | \\
-\mathrm{C} & - & \mathrm{C} & - & \mathrm{C} & - & \mathrm{C}- \\
| & & | & & | & & | \\
\mathrm{Cl} & & \mathrm{H} & & \mathrm{O} & & \mathrm{H} \\
 & & & & | & & \\
 & & & & \mathrm{C{=}O} & & \\
 & & & & | & & \\
 & & & & \mathrm{CH_3} & &
\end{array}\right]_m
$$

Each monomer imparts specific properties to the polymer. The vinyl chloride imparts hardness to the molecule, whereas the vinyl acetate imparts flexibility and solubility.

FUNCTIONALITY AND POLYMERIZATION MECHANISMS

The functionality of a molecule is the number of reactive groups it contains that can react with other molecules. The functionality of molecules determines the type of structure that can be formed.

The effects of functionality may be summarized as follows:

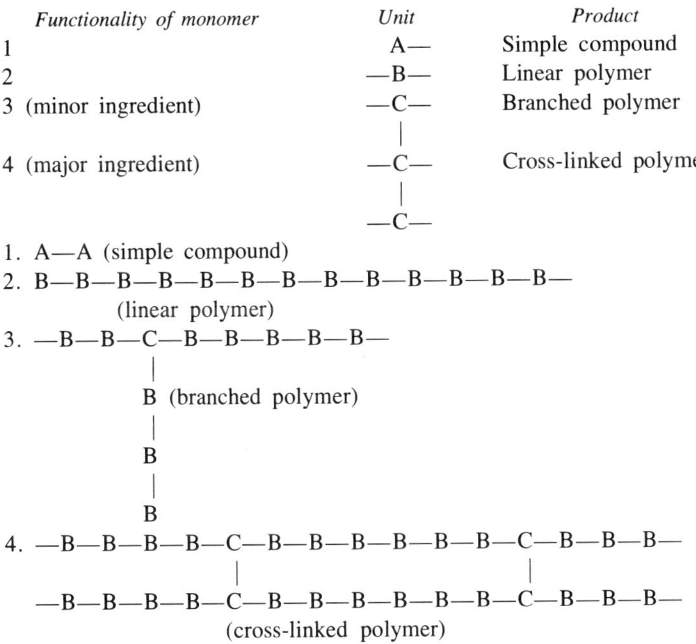

Functionality of monomer	*Unit*	*Product*
1	A—	Simple compound
2	—B—	Linear polymer
3 (minor ingredient)	—C— (with one branch bond)	Branched polymer
4 (major ingredient)	—C— (with two branch bonds)	Cross-linked polymer

TYPES OF POLYMERIZATION

Addition Polymerization

Many polymers are formed from compounds that contain carbon-to-carbon double bonds by a process that is called addition polymerization.

For example, ethylene polymerizes upon heating (200–275°F) under high pressure (1000 atm) to form polyethylene:

```
H  H    H  H     H  H  H  H
|  |    |  |     |  |  |  |
C==C  + C==C  →  —C—C—C—C———————
|  |    |  |     |  |  |  |
H  H    H  H     H  H  H  H
```

This process continued by the successive addition of $CH_2{=}CH_2$ molecules until chains of hundreds of $—CH_2—CH_2—$ units are produced.

Polymers formed from a single monomer are called homopolymers.

Condensation Polymerization

This type of polymerization occurs by reaction between molecules with the elimination of small molecules such as water.

Polyesters are of this type:

```
     H  H                 O  H  H  H  H  O
     |  |                 ||  |  |  |  |  ||
HO—C—C—OH           HO—C—C—C—C—C—C—OH →
     |  |                    |  |  |  |
     H  H                    H  H  H  H

  Ethylene glycol            Adipic acid
```

```
[      H H      O  H H H H  O       H H      O  H H H H  O        ]
[      | |      ||  | | | |  ||      | |      ||  | | | |  ||       ]
[ HO—C—C—O—C— C—C—C—C —C—O— C—C—O—C— C—C—C—C —C—OH ]  + nH2O
[      | |         | | | |          | |         | | | |           ]
[      H H         H H H H          H H         H H H H           ]
```

Condensation polymers used in coatings are polyesters, alkyds, phenol-formaldehyde, urea-formaldehyde, melamine-formaldehyde, and so on.

METHODS OF POLYMERIZATION

There are four different methods of polymerization in general use: bulk, solution, emulsion, and suspension.

Bulk Polymerization

In this method the catalyst and initiator are added to the monomer; as the polymerization proceeds, the viscosity increases until the mass becomes too viscous to agitate. Most polymerizations are exothermic, giving off a large amount of heat which makes the reaction difficult to control. Bulk polymerization is used in the fusion process for alkyds and bodied oils and for cast plastics in which the time element is relatively unimportant so that low temperatures may be used.

Solution Polymerization

In the solution process the monomers plus initiator and catalyst are dissolved in an inert liquid which may or may not be a solvent for the polymer. If the polymer is insoluble in the medium, it precipitates out as it is formed. If the polymer is soluble, it is precipitated with a nonsolvent.

The solvent polymerization process is more easily controlled than the bulk process, and a more uniform product is obtained. The lower viscosity makes heat transfer and handling of the reaction mixture much easier. This method is used for both condensation polymers; (alkyds, phenolics, epoxies, etc.) and addition polymers; (vinyls, acrylics, etc.).

Emulsion Polymerization

The monomer, which is insoluble in water, is emulsified with a surfactant. The polymerization proceeds very rapidly in the presence of an initiator, yielding a high-molecular-weight product in finely divided form. Excellent control of the process is possible, and a very uniform polymer is obtained. Because the polymer is obtained in the dispersed phase, there is no appreciable increase in the viscosity of the reaction mixture as the polymerization proceeds. This process may be used for addition polymerization.

Suspension Polymerization

In suspension polymerization the monomer is dropped into water and polymerized. However, no emulsifying agent is used, and the monomer droplets are much larger than those in an emulsion polymerization. The mechanism of suspension polymerization is similar to that of the bulk method. A thickener is added to the water to keep the polymer beads in suspension.

MOLECULAR WEIGHT EFFECTS

As the molecular weight of a polymer increases, the physical properties increase or generally improve. Below molecular weights of 1,000 we find solvents, plasticizers, and oils. In the range of 1,000 to 10,000 we find very low-molecular-weight polymers, bodied drying oils, alkyds, polyesters, and so on. In the range from 10,000 to 1,000,000 we find the true or higher polymers such as cellulosics, rubber, vinyls, and so forth. As the molecular weight increases, the physical properties such as tensile strength increase up to a molecular weight of about 150,000. Beyond this point there is very little in the physical properties. As the molecular weight increases, the solubility of the resin decreases so that it becomes more difficult to use high-molecular-weight resins in solution. The solutions have very high viscosities at low

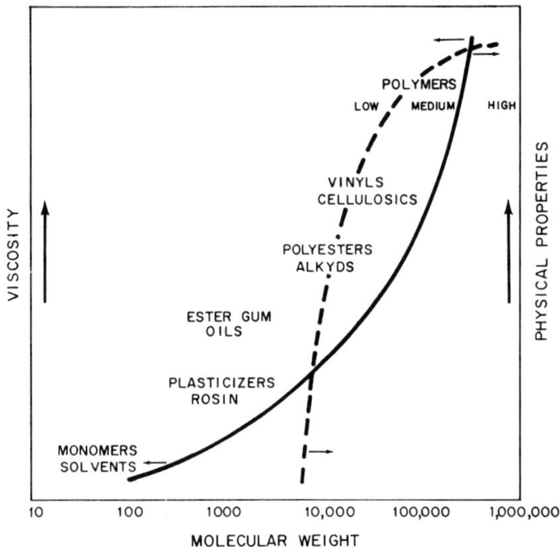

FIGURE 5.1. General properties vs. molecular weight.

solids content and require very strong solvents. (See Figure 5.1.) However, in dispersion systems such as latex paints it is possible to use high-molecular-weight polymers without these restrictions. Here the polymer is dispersed in water, and the viscosity is entirely dependent on the viscosity of the external water phase.

CLASSES OF RESINS

Acrylics

Acrylic resins are very versatile resins for use in coatings. While relatively high in cost, they are noted for their good color retention and exterior durability. They are available in a wide range of flexibilities, in both latex and water-soluble form. (See Tables 5.2 and 5.3.)

Acrylics are polymers of esters of acrylic acids:

$$CH_2{=}\overset{H}{C}{-}COOH$$

Acrylic acid

$$CH_2{=}\overset{CH_3}{\overset{|}{C}}{-}COOH$$

Methacrylic acid

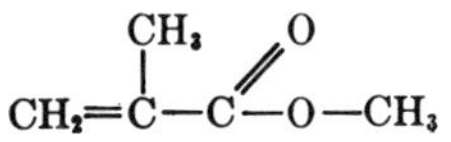

$$CH_2{=}C(CH_3){-}C({=}O){-}O{-}CH_3$$

Methyl methacrylate

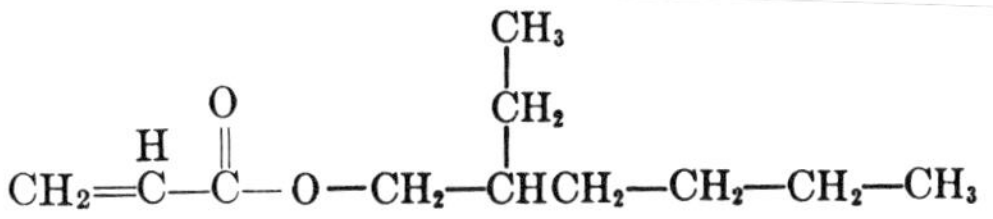

$$CH_2{=}CH{-}C({=}O){-}O{-}CH_2{-}CH(CH_2CH_3)CH_2{-}CH_2{-}CH_2{-}CH_3$$

Ethyl-hexyl-acrylate

The polymers are formed by addition polymerization:

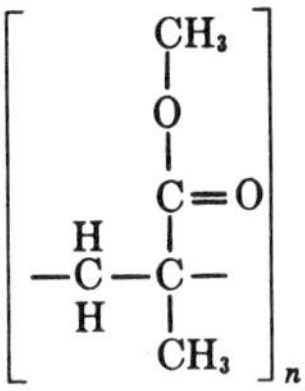

$$\left[-CH_2-C(CH_3)(C({=}O){-}O{-}CH_3)- \right]_n$$

Methyl methacrylate polymer

Various copolymers of these materials can be made. Methyl methacrylate produces a relatively hard polymer, while ethyl acrylate and ethylhexyl-acrylate are relatively soft materials and are included as a plasticizing monomer. Some acrylic acid may be copolymerized in a latex to improve adhesion. Polyacrylic acid in the form of an ammonia salt is water-soluble and can be used as a thickener. Water-soluble acrylic resins are made by incorporating acrylic acid into the molecule and then solubilizing with an amine. Acrylics can be made in the form of thermosetting materials which cross-link on baking. Copolymers of acrylonitrile and butadiene produce soft, flexible polymers.

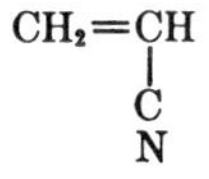

$$CH_2{=}CH{-}C{\equiv}N$$

Acrylonitrile

Alkyds

Alkyds are a group of resins that can be called oil-modified polyester resins. They are polymers formed by the reaction of polybasic acids, polyhydric alcohols, and fatty monobasic acids. The polybasic acid may be phthalic anhydride, maleic anhydride, and so on. The polyhydric alcohol may be glycerin, pentaerythritol, and so on, and the fatty acids are from soya, linseed, and so forth. A typical alkyd has the following structure, where R = fatty acid chain:

TABLE 5.2. ACRYLIC POLYMERS—EMULSIONS.

Trade name	Producer	NVM % Wt.	NVM % Vol.	pH	Pounds per gal.	Viscosity, poises	Particle size, microns	MFT, °F	Remarks
Arolon[R]									
801-W44	Ashland	44	41	7.7	8.8	—	—	—	
X820-W49	Ashland	49	46	8.5	8.8	1.5–4.0	—	—	—
X840-W40	Ashland	40	37	7.5	8.7	0.5–2.0	—	—	—
Durex									
X-400	Grace	53	—	6.0	8.8	0.15	—	—	Self-curing, solvent-resistant, soft film.
X-410	Grace	46	—	6.0	8.7	0.15	—	—	Self-curing, very soft film.
X-411	Grace	46	—	6.0	8.8	0.15	—	—	Self-curing, hard film.
X-441	Grace	50	—	3.0	8.8	0.15	—	—	Cross-linking, hard film, superior solvent resistance.
NeoCryl[R]									
A-601	Polyvinyl Chemical Ind.	32	28	8.0	8.4	1.0	—	< 32	General purpose, excellent adhesion to plastics.
A-602	Polyvinyl Chemical Ind.	32	29	8.0	8.6	1.0	—	55	For metal coatings harder than A-601.
A-603A	Polyvinyl Chemical Ind.	36	—	8.0	8.4	1.0	—	< 32	Base for trade sales varnish.
A-604	Polyvinyl Chemical Ind.	32	28	8.0	8.5	2.0	—	63	High quality wood finishes.
A-610	Polyvinyl Chemical Ind.	42	—	7.5	8.6	5.0	—	30	Uncoalesced for formulation latitude.

A-620	Polyvinyl Chemical Ind.	40	39	8.0	8.5	2.5	—	—	For hard, high-gloss industrial coatings.
A-621	Polyvinyl Chemical Ind.	40	39	8.5	8.5	6.0	—	—	Low cost, fast dry, flash-rust-resistant, high gloss.
A-622	Polyvinyl Chemical Ind.	32	29	8.2	8.5	3.0	—	—	Humidity resistance, water resistance finish.
A-624	Polyvinyl Chemical Ind.	40	39	8.5	8.5	2.5	—	—	Low cost, mar-resistant wood finishes.
A-630	Polyvinyl Chemical Ind.	44	43	8.0	8.6	6.0	—	—	Uncoalesced for formulation latitude.
BT-20	Polyvinyl Chemical Ind.	40	—	5.0	8.9	0.2	—	10	Cross-linkable.
BT-24	Polyvinyl Chemical Ind.	45	—	5.0	8.9	0.3	—	14	High-gloss bake enamels.
BT-175	Polyvinyl Chemical Ind.	40	—	4.0	8.8	1.0	—	24	Cross-linkable pigment dispersant.
A-1031	Polyvinyl Chemical Ind.	46	—	4.5	8.7	3.0	—	< 0	Soft, self-cross-linking.
A-1038	Polyvinyl Chemical Ind.	46	—	5.0	8.9	3.0	—	6	Self-cross-linking.
Synthemul[R]									
97-375	Reichhold Chemicals Inc.	50	47	9.0	8.8	1.1	0.2–0.5	48	Internal/exterior latex flat/semigloss paints.
40-422	Reichhold Chemicals Inc.	49	47	8.6	8.6	1.0 max	—	65	Latex maintenance coatings.
90-757	Reichhold Chemicals Inc.	40	36	8.5	8.9	0.5 max	0.05–0.15	95	Air dry enamels.
90-659	Reichhold Chemicals Inc.	49	47	8.5	8.6	1.0	—	—	Baking enamels.

Trade name	Producer	NVM % Wt.	NVM % Vol.	pH	Pounds per gal.	Viscosity, poises	Particle size, microns	MFT, °F	Remarks
Rhoplex[R]									
AC-19	Rohm & Haas Co.	44	40	9.5	8.8	—	—	46	Low-cost interior flat paints.
AC-22	Rohm & Haas Co.	44	41	9.8	8.9	—	—	46	Interior flat and semi-gloss paints.
AC-25	Rohm & Haas Co.	46	42	9.3	8.8	—	—	64	Interior paints with flow and leveling.
AC-33	Rohm & Haas Co.	46	43	9.6	8.9	—	—	46	Exterior house paints.
AC-61	Rohm & Haas Co.	46	43	9.8	8.9	—	—	64	Harder-floor paints.
AC-64	Rohm & Haas Co.	60	56	9.3	8.8	—	—	45	Exterior paints with adhesion to chalk.
AC-73	Rohm & Haas Co.	46	42	9.5	8.9	—	—	95	Harder; modifier.
AC-234	Rohm & Haas Co.	46	43	9.8	8.9	—	—	54	Exterior house paints.
AC-235	Rohm & Haas Co.	46	43	9.5	8.8	—	—	48	Exterior house paints.
AC-388	Rohm & Haas Co.	50	47	9.5	8.8	—	—	50	Universal vehicle.
AC-490	Rohm & Haas Co.	46	42	9.3	8.8	—	—	64	Interior gloss paints.
AC-507	Rohm & Haas Co.	46	43	9.3	8.8	—	—	57	Exterior gloss paints.
AC-604	Rohm & Haas Co.	46	—	10.0	9.0	—	—	—	Thermosettng exterior, hard.
AC-658	Rohm & Haas Co.	47	—	10.0	8.9	—	—	—	Thermosetting exterior, soft.
AC-707	Rohm & Haas Co.	65	63	9.4	9.0	—	—	45	High solids version of Rhoplex AC-388.
AC-1024	Rohm & Haas Co.	50	—	4.0	8.7	—	—	—	Fast cure DRC clear and topcoats, conversion varnishes.

AC-1025	Rohm & Haas Co.	42	39	9.4	8.9	—	—	—	DRC basecoats.
B-85	Rohm & Haas Co.	38	35	9.8	8.9	—	—	> 194	Extremely hard.
C-72	Rohm & Haas Co.	45	41	9.8	8.9	—	—	95	Hardening modifier.
E-32	Rohm & Haas Co.	46	—	3.2	8.8	—	—	< 32	Thermosetting, soft.
HA-8	Rohm & Haas Co.	46	—	3.0	8.7	—	—	< 32	Thermosetting, flexible.
HA-12	Rohm & Haas Co.	45	—	3.0	8.8	—	—	41	Thermosetting harder than HA-8.
HA-16	Rohm & Haas Co.	45	—	3.0	8.8	—	—	72	Thermosetting, rigid, modifier.
MV-1	Rohm & Haas Co.	46	43	9.5	8.9	—	—	46	Maintenance paint over metal.
MV-2	Rohm & Haas Co.	46	43	9.5	8.9	—	—	46	Flash-rust-resistant maintenance paints.
MV-9	Rohm & Haas Co.	45	43	9.0	8.7	—	—	50	Resistant to solvents, chemicals, humidity, flash rusting.
Rhoplex[R]									
MV-17	Rohm & Haas Co.	45	43	8.3	8.7	—	—	50	Gloss topcoats over metal.
MV-23	Rohm & Haas Co.	43	—	9.7	8.8	500–2000	—	46	Metal primers and stain blocking sealer.
SS-521	Rohm & Haas Co.	50	—	9.8	8.9	—	—	113	Clear coatings for concrete and masonry.
WL-81	Rohm & Haas Co.	41	—	7.5	8.7	—	—	135	Clear and pigmented over air dry industrial lacquers.
WL-91	Rohm & Haas Co.	41	39	7.5	8.6	500–2000	—	135	Clear and pigmented, gasoline resistant industrial air dry lacquer.

Trade name	Producer	NVM % Wt.	NVM % Vol.	pH	Pounds per gal.	Viscosity, poises	Particle size, microns	MFT, °F	Remarks
Acrysol[R]									
I-94	Rohm & Haas Co.	30	—	4.0	8.6	—	—	80	Flat pH/viscosity curve, good pigment dispersion.
I-98	Rohm & Haas Co.	30	—	4.0	8.6	100	—	78	Low foaming, good pigment dispersion.
WS-12	Rohm & Haas Co.	30	—	8.7	8.7	—	—	50	
WS-24	Rohm & Haas Co.	36	—	7.4	8.7	—	—	50	Can be water-solubilized; forms glossy, water-resistant film.
WS-32	Rohm & Haas Co.	22	—	8.2	8.6	—	—	41	Similar to Acrysol WS-24 but softer.
WS-50	Rohm & Haas Co.	38	—	7.6	8.8	—	—	50	Similar to Acrysol WS-24 but harder.
WS-68	Rohm & Haas Co.	38	—	7.5	8.7	500	—	< 41	Industrial baking enamels with outstanding spray characteristics.
Ucar									
380	UCC	48	45	6.5	8.8	5.0	0.2–0.7	48	Interior/exterior gloss paints.

503	UCC	58	54	4.5	9.2	10.0	0.4	57	Universal paint vehicle.
508	UCC	53	49	4.5	9.1	2.0	0.6	54	Exterior house and trim paints, interior semigloss high quality.
4312	UCC	46	42	8.7	8.7	0.8	0.15	65	Strippable coatings.
4341	UCC	46	45	8.7	8.6	0.4	0.2	36	Medium-duty maintenance and industrial finishes.
4358	UCC	45	43	8.6	8.7	0.9	0.15	70	Heavy-duty maintenance and industrial finishes.
4362	UCC	48	45	6.8	8.8	4.0	0.6	46	Light duty, gloss maintenance coatings.
4414	UCC	45	—	8.7	8.7	0.8	0.2	65	Tough, flexible, strippable coating.
4426	UCC	45	—	8.7	8.6	0.4	0.2	46	Flexible: excellent for primers.
4433	UCC	47	—	8.7	8.7	0.8	0.2	90	Hard, glossy, industrial lacquers.
4510	UCC	43	40	8.0	8.8	2.0	0.2	96	High-gloss baking enamels.
4550	UCC	45	42	4.7	8.9	1.5	0.3	80	Hard, self-cross-linking for industrial finishes.
4620	UCC	45	43	8.0	8.8	2.5	0.3	59	Cross-linking, coil coatings.

TABLE 5.3. ACRYLIC WATER-REDUCIBLE RESINS.

Trade name	Producer[a]	NVM % Wt.	NVM % Vol.	Cosolvent[b]	Solubilizer	Visc., Gardner Holdt	Color, Gardner Max.	pH	Pounds per gal.	Acid number solids	Remarks
XC-4011	American Cyanamid	75	70	BuOH		Z_5–Z_7	1	—	8.5	100	Anionic resin for ED and W/R finishes.
XC-4012	American Cyanamid	48	44	EGMEE				—	8.7	100	
Arolon 557-D70	Ashland	70	56	IpOH		Z_5–Z_6	2	—	8.4	75	Industrial baking enamel.
7415	Cargill	70	65	BuOH		Z_5–Z_6	8	—	8.2	55	High-gloss, air dry industrial coating.
311-203	Conchemco	55	48	EGMEE/ NH_4OH IpOH		Z–Z_2	1	—	8.7	80	
Coroc Resin A1744-V3	Cook Paint & Varnish Co.	75	72	BuOH		Z–6	1	—	8.5	100	Industrial baking enamel.
Z2588-V8	Cook Paint & Varnish Co.	80	77	EGMBE		Z_4–Z_6	1	—	8.7	66	Industrial baking enamel.
Synthemul 90-587	RCI	65	56	PrOH	Amines	Z_5	1	—	8.4	77	Baking enamels.
90-588	RCI	70	67	EGMBE	Amines	Z_6	1	—	8.4	77	Dipping, flow enamels.
90-702	RCI	65	63	EGMBE	Amines	Z_6	1	—	8.5	77	Baking enamels.
90-734	RCI	75	73	EGMBE PrOH	Amines	Z_8	6	—	8.9	115	Primer, coil coatings.
97-378	RCI	72	68	EGMBE	Amines	Z_6	4	—	8.8	80	Coil coating.
Acryloid WR-97	R&H	70		IpOH EGMBE	Amines	Z_6	< 1	—	8.3	39	Industrial finishes cross-link with amino resins.

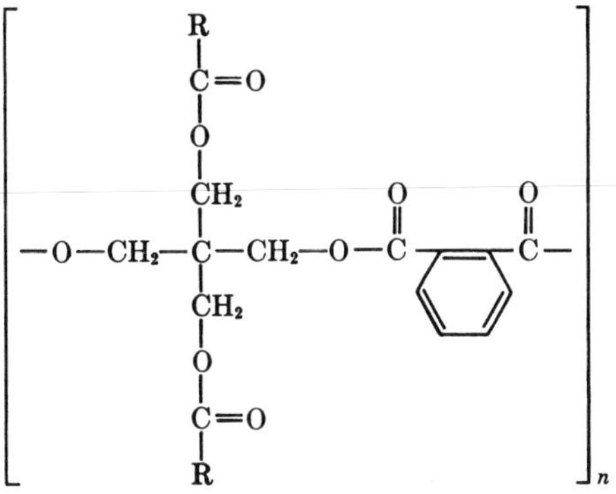

Alkyd resins are formed by a condensation reaction with the elimination of water; they are then postemulsified. The greater the oil content of an alkyd is, the greater the flexibility. Alkyds have good adhesion and good compatibility with other resins; they are sometimes added to latex paints to improve adhesion. In emulsion form alkyds sometimes show loss of dry on prolonged storage.

Alkyds can be produced in a water-soluble form by formulating a low-molecular-weight alkyd with a high proportion of ether and alcohol groups. A low-molecular-weight alkyd containing a high proportion of carboxyl groups can be made water-soluble by forming the ammonia salt.

Another method of producing a water-soluble alkyd is to incorporate a trifunctional acid such as trimellitic anhydride into the molecule.

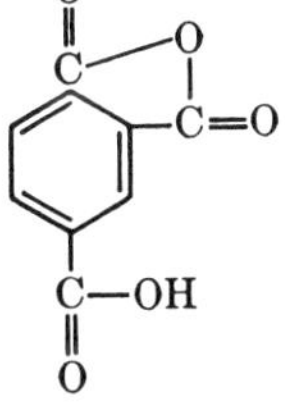

Trimellitic anhydride

The selection of the polyol as well as the polybasic acid affects the hydrolysis rate of the polyester. The number of ester groups adjacent to residual free hydroxyl plays an important part in determining the initial hydrolysis rate. Listed below are polyols in order of increasing hydrolysis resistance:

Polyol	Carbons between OH	Functionality	OH Primary	OH Secondary
Glycerin	2	3	2	1
Ethylene glycol	2	2	2	0
Propylene glycol	2	2	1	1
DiPE	3	6	6	0
PE	3	4	4	0
TME	3	3	3	0
TMP	3	3	3	0
Neopentyl glycol	3	2	2	0
DMPA	3	2	2	0
TMPD	3	2	1	1
Diethylene glycol	4	2	2	0
Dipropylene glycol	4	2	0	2
Triethylene glycol	6	2	2	0
CHDM	6	2	2	0
HBPA hydrogenated bisphenol	9	2	0	2

Manufacture of Water-reducible Alkyds

Many paint companies that have made solvent alkyds for a long time are now making water-reducible alkyds. (See Table 5.4.) The equipment is essentially the same, that is, Dowtherm heated kettles fitted with condensers.

The following water-soluble alkyd formula and processing yield the resin properties described:

Water-Soluble Alkyd Formula

	Parts by weight
Linseed fatty acids	357
Trimethylol propane	335
85% Isophthalic acid[a]	299
Trimellitic anyhdride	99
	1090
Less water of reaction	−90
	1000

Processing. Charge fatty acids, trimethylol propane, and isophthalic acid into reaction kettle. Heat over a three-hour period to 460°F (238°C), and hold for acid value below 10. Cool to 360°F (182°C), add TMA, and hold at 340–350°F (171–177°C) for 50–65 acid number and 20-second cure[a] Amoco IPA 85. Cool to 310°F (154°C). Thin to 75% solids with cosolvent. Cooking time is 14 hours.

Resin Properties

Acid number (solids)	50
Hydroxyl number (solids)	230–240
Viscosity (Gardner-Holdt)	Z_6
NVM %	75
Volatile	1:1 Butanol/butyl cellosolve
Cure[a] 200°C	20 seconds

[a]Cure time required from liquid to gel of thin film on hot plate.

Amino Resins

These resins include urea-formaldehyde, melamine-formaldehyde, and so on. Amino resins are not used by themselves in coatings, but are used to cross-link alkyd resins, polyester resins, acrylic resins, epoxy resins, and so forth. Amino resins require baking to attain curing. To make amino resins such as urea-formaldehyde and melamine-formaldehyde water-soluble, the butyl alcohol group is replaced with methyl or ethyl alcohol.

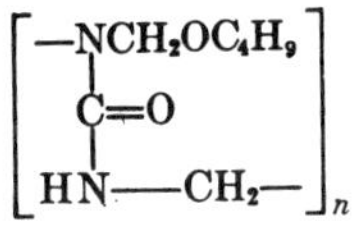

$$\left[\begin{array}{l} -\text{NCH}_2\text{OC}_4\text{H}_9 \\ \;\;\text{C}{=}\text{O} \\ \text{HN}{-}{-}\text{CH}_2{-} \end{array}\right]_n$$

Urea-formaldehyde (butylated)

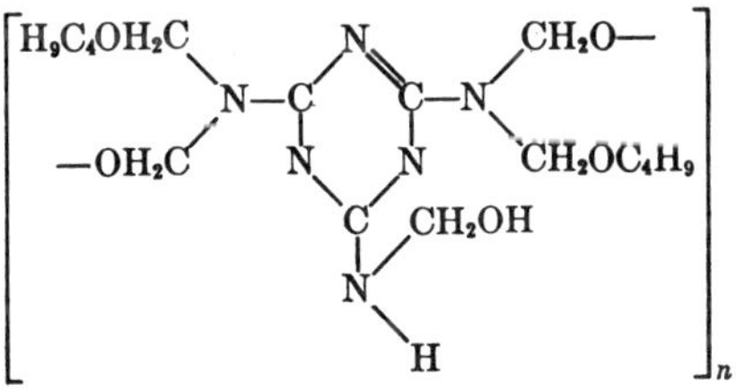

$H_9C_4OH_2C$, $-OH_2C$ (on N); N–C (triazine ring: N, C, N, C, N) C–N; CH_2O-, $CH_2OC_4H_9$ (on N); C–N(H)CH_2OH — $[\ ]_n$

Melamine-formaldehyde (butylated)

Amino resins cross-link with resins containing hydroxyl and carboxyl groups with the elimination of water.

Table 5.5 lists water-soluble amino resins.

Bitumens

The bitumens are basically divided into petroleum and natural asphalts and coal tars.

Asphalt has excellent resistance to sunlight, weathering exposures and summer heat, whereas coal tar pitches tend to flow and alligator in such exposures to an extent that their use is not recommended for direct weather-

TABLE 5.4. ALKYDS—WATER-REDUCIBLE.

Trade name	Producer	NVM % Wt.	NVM % Vol.	Cosolvent	Solubilizer	Visc., Gardner Holdt	Color, Gardner	pH	Pounds per gal.	Acid value solids	Type of oil	Remarks
Arolon 332-Z70	Ashland	70	63	PrOH EGMBE		Z_2–Z_3	6		8.9	60	Saff	Water dispersion Code 542-150.
363-WA8-50	Ashland	50	44	EGMBE	TEA	S–X	6	7.1	8.7	48	Soya	Code 542-050.
369-B80	Ashland	80	74	BuOH		Z_3–Z_5	10		8.8		Soya	Water dispersion type Code 542-055.
376-WA8	Ashland	50	44	*t*-BuOH EGMBE	TEA	S–W	5	7.2	8.7		Saff	Code 542-070.
378-WA8-50	Ashland	50	42	*t*-BuOH/ EGMBE		Z–Z_2	7		8.5	40	Soya	Water-reducible alkyd.
580-W42	Ashland	42	39		TEA	7.25 psi	Milky	7.5	8.9		Saff	Dispersion type Code 543-030.
585-W43	Ashland	43	38		TEA	3–15 psi	Milky	7.4	8.9		Saff	Dispersion type Code 543-040.
969-A8-70	Ashland	70	66	EGMBE		Z_6–Z_7	18		8.8	75	Tall	Copolymer disp. type Code 543-070.
7404-70	Cargill	70	61	BuOH		Z_3	3		8.8		Soya	Exempt cosolvent excellent hard, low temp.
7407	Cargill	80	76	BuOH		Z_4–Z_6			8.7		Soya	For air dry industrial coatings.
311-203	Conchemco	55	48	EGMEE/ IpOH	NH_4OH	Z–Z_2	1	8.9	8.7	80		Thermoset: crosslink w/ amine or epoxy.

Coroc L99-V8	CPV	70	66	EGMBE		U–X	7	8.5	45	Lins	Industrial baking enamel.
Z1055-M	CPV	75	72	PGGME/ 2-BuOH		Z_2–Z_3	6	8.6	59	Saff	Air dry high-gloss enamel.
Z-1055-V8	CPV	73	69	EGMBE		Z_3–Z_4	6	8.7	57	Saff	Industrial baking enamel.
T-1070M	CPV	75	72	BuOH/ PGMME		Z_2–Z_4	6	8.6	59	Tall	Industrial baking enamel.
Z-2055M	CPV	70	66	PGMME		Z_1–Z_2	6	8.6	59	Saff	Fast air dry enamel.
Z-2055-V8	CPV	70	65	EGMBE		Z_2–Z_3	5	8.7	59	Saff	Industrial baking enamel.
B-153-90	Koppers	75	70	EGMBE	base	Z_3–Z_5	12	8.9	57	Lin	Epoxy alkyd, air dry and thermoset.
B-153-151	Koppers	75	70	EGMBE	base	Z_4–Z_6	7	8.9	58	Soya	Epoxy alkyd, air dry and thermoset.
B-153-120	Koppers	75	70	EGMBE	base	X–Z_1	3	8.9	63	Syn	Roller coat.
B-414-66	Koppers	70	63	EGMBE	base	Z_2–Z_4	8	8.7	57	Soya	Forced air dry, high gloss.
B-406-122	Koppers	70	64	EGMBE	base	Z_4–Z_6	2	9.0	59		Good color retention and chemical resistance.
Aquamac[R] 1070	McWhorter	70	61	IbOH	Amine or ammonia	Z_5–Z_6	11	8.4	70	Tung	Water-soluble oxalone baking finish.
1075	McWhorter	70	61	IbOH	Amine or ammonia	Z_5–Z_6	8	8.6	60	Tall	Bake or air dry finishes.
1085	McWhorter	70	61	IbOH	Amine or ammonia	Z_5–Z_7	8	8.6	60	Tall	Air dry or form dry universal type.

Trade name	Producer	NVM % Wt.	NVM % Vol.	Cosolvent	Solubilizer	Visc., Gardner Holdt	Color, Gardner	pH	Pounds per gal.	Acid value solids	Type of oil	Remarks
1100	McWhorter	70	61	EB/EGMBE/ IbOH		Z_3–Z_6	6		8.9	50		Acrylic fast air dry high gloss enamel.
1200	McWhorter	70	61	EB/EGMBE/ IbOH		Z_2–Z_5	9		8.6	50		Urethane mod. high corrosion resistant.
1300	McWhorter	70	61	EGMBE/ IbOH		Z–Z_3	6		8.9	55		Low cost oil-free polyester, high-quality baking finish.
1400	McWhorter	70	61	EGMBE	Amine	Z_3–Z_4	8		8.4	55		Epoxy ester, optimum corrosion and chemical resistance.
Beckosol[R] 13-403	RCI	75	72	EGMBE	None	Z_1–Z_3	8		8.7	58		Baking primers and enamels.
13-412	RCI	80	76	EGMBE/ PrOH	None	Z_3–Z_5	11		8.8	45		Baking primer.
13-418	RCI	75	71	EGMBE/ EGMEE	None	Z_4–Z_6	3		9.3	60		Baking enamel.
13-420	RCI	50	45		None	W	4	8.5	8.6	120		Flow agent.
13-432	RCI	36	35		None	F–J	18		8.5			Baking primer.
13-436	RCI	70	67	EGMEE	None	Z_1–Z_3	8		8.5	65		Air dry and baking enamels.

13-488	RCI	75	72	EGMBE	None	Z_4–Z_6	11	8.9	62		Air dry and force dry primers and enamels.
92-188	RCI	65	57	BuOH/ EGMBE	None	Z_1–Z_3	6	8.5	45		Air dry and baking enamel.
WR-2301 BC-75	Reliance	75	71	EGMBE/ BuOH	None	Z_1–Z_3	7	8.5		F.A.	Oxidizing
WR-3101 BC-75	Reliance	75	69	EGMBE/ BuOH	None	Z_4–Z_5	6	8.6		F.A.	Nonoxidizing
WR-4001 BC-70	Reliance	70	64	EGMBE/ BuOH	None	Z_1–Z_3	8	8.6		F.A.	Fast air dry
WR-5001 BC-75	Reliance	75	70	EGMBE/ BuOH	None	U–Z_2	9	8.8		F.A.	Primer vehicle
Kelsol[R] 3900	Spencer Kellogg	75	69	EGMBE/ BuOH		Z_6	6	8.8	54		Very fast, air dry vehicle for prime and topcoat.
3902	Spencer Kellogg	75	69	EGMBE/ BuOH		Z_6	6	8.6	32		Fast air dry vehicle with excellent corrosion resistance.
3903	Spencer Kellogg	75	70	EGMBE/ BuOH		Z_6	6	8.5	32		Air dry vehicle with high flexibility.
3904	Spencer Kellogg	75	69	EGMBE/ BuOH		Z_6	6	8.7	32		Fast, air dry, low cost, good corrosion resistance.

TABLE 5.5. WATER-SOLUBLE AMINO RESINS.

Trade name	Producer	NVM % Wt.	NVM % Vol.	Solvent	Viscosity, Gardner Holdt	Color, Gardner	Av. solids	Pounds per gal.	Notes
Beetle[R] 55	Amer. Cy.	75	70	Water	$X–Z_1$	1	2	9.9	Urea-methylated.
60	Amer. Cy.	88	82	IPA	V–Y	1	2	9.8	Urea—methylated, fast cure.
65	Amer. Cy.	100	100	IPA	$Z_3–Z_6$	1	2	9.8	Urea-methylated, fast cure.
Resimene[R] X-2970	Monsanto	90	87	Water	$X–Z_2$	1	—	10.3	Urea—fast cure.
Beckamine[R] 21-675	RCI	75	62	IPA	$Z–Z_4$	1	—	9.9	Urea—water-reducible.
Cymel[R] 373	Amer. Cy.	85	81	Water	$Z–Z_4$	1	—	10.5	Melamine—water-soluble.
385	Amer. Cy.	80	95	Water	U–W	1	—	10.3	Melamine—methylated.
1172	Amer. Cy.	45	33	Water	—	1	1	10.1	Melamine—tetramethylol glycoluril.
Resimene[R] X 707	Monsanto	80	75	Water	V–Y	1	—	10.6	Melamine—methylated, very rapid cure.
X 712	Monsanto	80	75	Water	V–Y	1	—	10.5	Melamine—methylated, very rapid cure.
714	Monsanto	80	75	Water	L–V	1	—	10.5	Melamine—methylated, enamel solubility.
X-2-720	Monsanto	90	87	Water	—	1	—	10.5	Melamine—methylated, rapid cure with good stability.
X-2-735	Monsanto	90	87	Water	$Z_1–Z_6$	1	—	10.5	Melamine—methylated, fast cure.
X-2-740	Monsanto	90	88	Water	$Z–Z_4$	1	—	10.3	Melamine—methylated, versatile.

ing. Coal tar pitches, on the other hand, have greater moisture resistance than the asphalts generally formulated for industrial coatings: the usual asphalt coatings will tend to lose adhesion to the substrate on continuous water or wet-soil immersion.

Thus basically the asphalts should be used in weathering exposures above ground and the coal tar pitches should be employed in preference to asphalt for continuous wet or damp exposures, water immersion, and soil burial.

Bitumens are available in emulsion form and are used for asphalt drive coatings, roof coatings, tank coatings, and so on.

Cellulosics

The chief sources of cellulose for the manufacture of cellulosic coating materials are wood, cotton, and cotton linters. Cellulose itself is insoluble except in a few specialized reagents (e.g., cuprammonium solution). It can be converted by suitable chemical treatment, involving esterification or etherification of the hydroxyl groups into soluble cellulosic derivatives.

Cellulosics include organic soluble cellulose esters, such as cellulose nitrate, acetate, propionate, butyrate, and mixed esters, and ethers such as ethyl and benzyl cellulose.

Ethers such as methyl, carboxy methyl, and hydroxyethyl are water-soluble. They are covered in Chapter 7.

Because of their high molecular weights, cellulosic resins have outstanding toughness and fast-drying properties. A considerable amount of strong solvent is required to dissolve cellulosic materials even when used in emulsion forms. The cellulosic molecule is as follows:

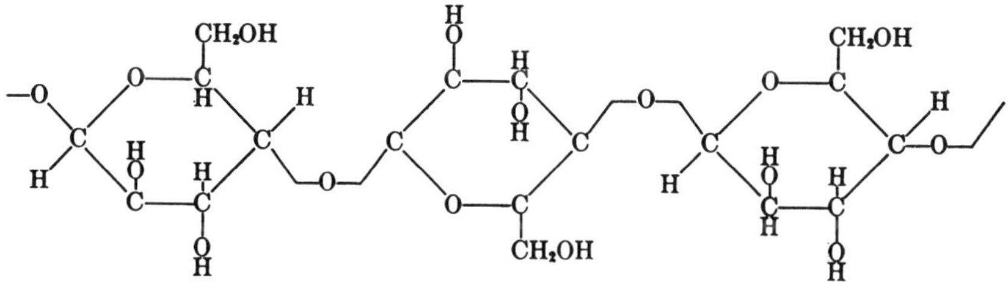

The hydroxyl group is replaced as follows:

Nitro	$—O—NO_2$
Ethyl	$—O—CH_2—CH_3$
Acetate-butyrate	$—O—C(=O)—CH_3$ $—O—C(=O)—CH_2—CH_2—CH_3$

Although the cellulose molecule contains three hydroxy groups per unit, the amount of substitution is usually less. For some applications the inflammability of nitrocellulose limits its use. Ethyl cellulose has outstanding alkali resistance.

Drying Oils

Vegetable and marine oils are triglycerides of long-chain unsaturated fatty acids containing small amounts of phosphatides, carbohydrates, and other impurities:

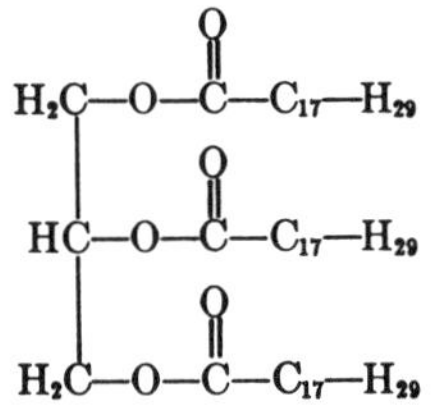

$$\begin{array}{l} H_2C-O-\overset{\overset{O}{\|}}{C}-C_{17}-H_{29} \\ \;| \\ HC-O-\overset{\overset{O}{\|}}{C}-C_{17}-H_{29} \\ \;| \\ H_2C-O-\overset{\overset{O}{\|}}{C}-C_{17}-H_{29} \end{array}$$

The majority of fatty acids have 18-carbon chains and vary in degree of unsaturation. Some of the common drying oils are linseed, safflower, dehydrated castor, and tung. Oils can be heat-bodied to a higher molecular weight.

A typical linseed oil emulsion made from a heat-bodied oil has the following properties:

Solids	60%
Viscosity	50–60 K.U.
pH	7.5–9.0
Particle size	$< 1.5\ \mu$

Oils combined with resins such as phenolics, maleics, and so on, are known as oleoresinous vehicles.

A water-soluble-type oil can be produced by forming the adduct between fumaric acid and tung oil as follows:

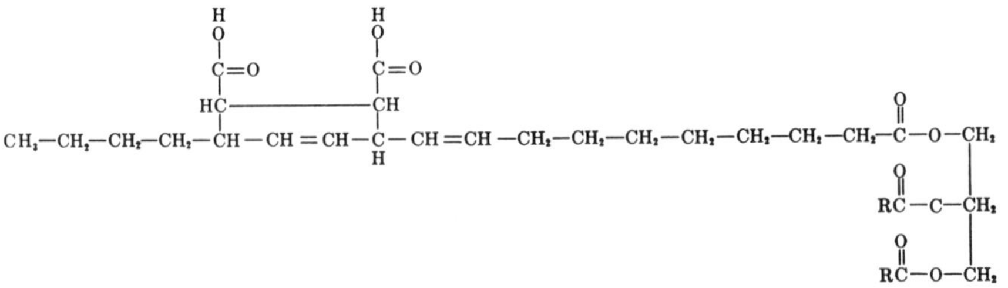

$$\begin{array}{l} \qquad\qquad\qquad\qquad H \qquad\qquad\quad H \\ \qquad\qquad\qquad\qquad O \qquad\qquad\quad O \\ \qquad\qquad\qquad\qquad C=O \qquad\quad C=O \\ \qquad\qquad\qquad\qquad HC\text{———————}CH \\ CH_3-CH_2-CH_2-CH_2-CH-CH=CH-\underset{H}{C}-CH=CH-CH_2-CH_2-CH_2-CH_2-CH_2-CH_2-CH_2-\overset{\overset{O}{\|}}{C}-O-CH_2 \\ \qquad\qquad\qquad\qquad\qquad\qquad\qquad\qquad\qquad\qquad\qquad\qquad\qquad\qquad\qquad\qquad\qquad \overset{\overset{O}{\|}}{RC}-C-CH_2 \\ \qquad\qquad\qquad\qquad\qquad\qquad\qquad\qquad\qquad\qquad\qquad\qquad\qquad\qquad\qquad\qquad\qquad \overset{\overset{O}{\|}}{RC}-O-CH_2 \end{array}$$

This oil is solubilized with ammonia and the addition of a cosolvent.

Epoxies

These resins are the condensation polymers of epichlorohydrin and bisphenol or their derivatives. The resin molecule contains both hydroxyl and epoxy functional groups. The typical structure is as follows:

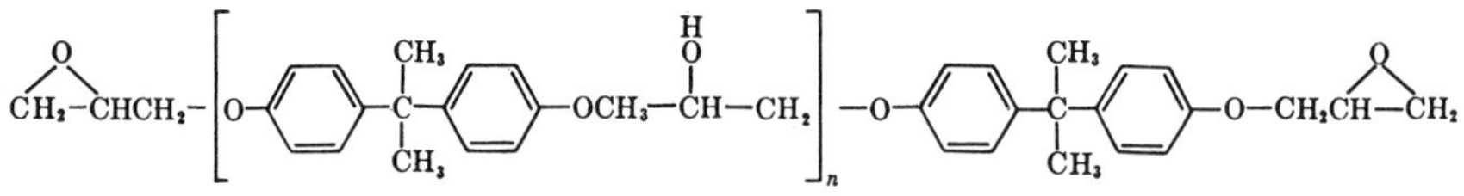

$$\overset{\text{O}}{\overbrace{CH_2-CH}}CH_2-\left[O-C_6H_4-\underset{CH_3}{\overset{CH_3}{\underset{|}{\overset{|}{C}}}}-C_6H_4-OCH_2-\overset{\overset{H}{O}}{\overset{|}{C}H}-CH_2\right]_n-O-C_6H_4-\underset{CH_3}{\overset{CH_3}{\underset{|}{\overset{|}{C}}}}-C_6H_4-O-CH_2\overset{\text{O}}{\overbrace{CH-CH_2}}$$

Epoxy resins can be polymerized through their reactive epoxy or hydroxyl groups using amines or polyamides. They can be esterified with fatty acids to produce air-dry coatings. The latter system is more adaptable for use in water-type coatings. These resins are commercially available in emulsion form, and are known for their adhesion, toughness, and chemical resistance. (See Table 5.6.)

Chlorinated Polyethers

These resins have good corrosion resistance at temperatures up to 240°F. The polymer has excellent water resistance, but because of discoloration a pure white cannot be produced. By use of a water-suspension system, coatings up to 40 mils in thickness can be applied in as few as two separate coats; after the water is evaporated, the resin is fused at temperatures of 425°F or higher.

The structure of the polymer is as follows:

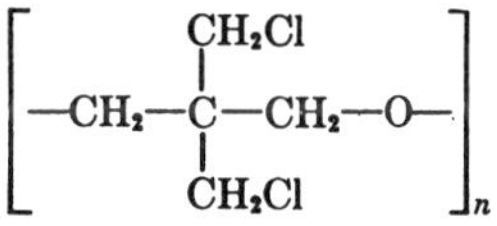

$$\left[-CH_2-\underset{CH_2Cl}{\overset{CH_2Cl}{\underset{|}{\overset{|}{C}}}}-CH_2-O-\right]_n$$

Fluorocarbons

The fluorocarbon resins are substituted ethylene polymers in which the hydrogen atoms have been replaced with fluorine or fluorine and chlorine atoms. The resulting polymers have outstanding chemical inertness, a low coefficient of friction, heat resistance (up to 450°F), and strength and flexibility over a wide temperature range.

Polytetrafluoroethylene and polychlorotrifluoroethylene are the most important resins of this type. They are available in dispersion form in water in some cases.

Hydrocarbon Resins

This large class includes terpene resins, coumarone-indene resins, and petroleum resins. The hydrocarbon resins have good acid and alkali resistance and are used principally to modify other resins.

Coumarone-indene resins are produced from coumarone and indene, which are obtained from coal tar naphthas. Commercial resins of this type are Picco(R)[a] and Cumar(R)[b].

Petroleum resins are obtained from the by-products from cracking petroleum oils. Commercial resins of this type are Velsicols(R)[c], Panarez(R)[d], and Picodiene(R)[e], and Lx[f].

Terpene resins are produced from compounds obtained from turpentine, which comes from pine trees. These are cyclic hydrocarbons with double bonds, such as dipentene, terpinolene, phillandrene, terpinene, and B pinene, which are polymerized into resins. Commercial products are Piccolyte(R)[g].

Phenolic Resins

These resins are produced from the condensation of various phenols with formaldehyde and derivatives. Substituted phenols are used to attain oil solubility for use with drying oils. Phenolic resins may yellow and darken on exposure. These resins have good durability and alkali resistance. A typical phenolic resin is:

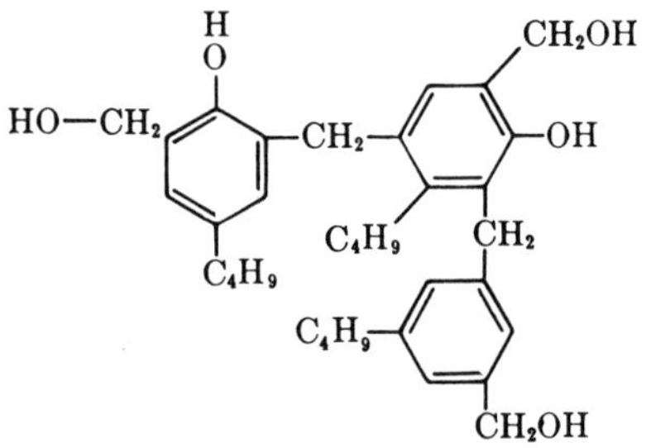

H O; CH_2OH; $HO—CH_2$; CH_2; OH; C_4H_9; CH_2; C_4H_9; C_4H_9; CH_2OH

Several water-soluble and water-dispersible phenolic resins are available commercially. (See Table 5.7.)

Polyamides

These resins are products of the reaction between polybasic acids and polyamines. They are used by themselves or as epoxy resin cross-linkers, and

[a]Pennsylvania Industrial Chemical Corp.; [b]Neville Chemical Company; [c]Vesicol Chemical Corp.; [d]Amoco Chemical Corp.; [e]Pennsylvania Industrial Chemical Corp.; [f]Neville Chemical Corp.; [g]Pennsylvania Industrial Chemical Corp.

TABLE 5.6. EPOXY RESINS—WATER-REDUCIBLE.

Trade name	Producer	NVM % Wt.	NVM % Vol	Cosolvent	Viscosity, Gardner Holdt	Color, Gardner	Pounds per gal.	Acid value	Remarks
Araldite[R] PR-805	Ciba Geigy	50	—	Xylene	28 cps	—	8.5		Epoxy ester emulsion for coatings.
PR-808	Ciba Geigy	60	—	EGMBE	150 poises	5	9.4		Epoxy ester solution for coatings—heat curing.
PR-9497	Ciba Geigy	60	—	EGMME	110 poises	3	9.5		Epoxy ester solution for coating—heat curing.
Epotuf[R]	RCI	70	66	EGMBE	Z_6–Z_7	7	8.4	57	Epoxy ester air dry primer and enamels.
Poly Tex[R] 611 Q	Celanese	—	60	—	80–100 poises	—	8.4	—	Epoxy ester emulsion used in water-red. baking primer.

TABLE 5.7. PURE PHENOLICS—WATER DISPERSIONS.

Trade name	Producer	NVM % Wt.	Solvent	Viscosity, poises @25°C	Pounds per gal.	Notes
BRUC-2660	UCC	41	Water	9–12	9.34	Heat-reactive—aqueous phenolic.
BRUC-2260	UCC	48	Water	12–18	9.34	Dispersion—compatible with many latexes.

have outstanding strength, toughness, and abrasion resistance. (See Table 5.8.)

A typical polymer is as follows:

$$\left[\overset{\displaystyle H}{\overset{|}{—N}}—CH_2—CH_2CH_2—CH_2—CH_2—\overset{\displaystyle O}{\overset{\|}{C}}— \right]_n$$

Polyesters

Polyester resins are polymers formed by condensation reaction between acids and alcohols. They require curing by heat with cross-linking agents such as amino resins, which react with their hydroxy and carboxy end groups. They have good color retention, abrasion resistance, and durability. They can be produced in water-soluble form and have found intensive industrial use such as coil coatings. (See Table 5.9.)

Polyethylene

This is a low-cost waxlike polymer which has good water resistance and slipperiness.

Polyethylene has the structure:

$$[—CH_2—CH_2—]_n$$

It is available in water dispersion form and added to coatings to impart scuff resistance and mar proofing.

Polyurethanes

The polyurethane resins, which are among the newest resins used in coatings, were developed in Germany during World War II. The fundamental urethane reaction is that of an isocyanate with an alcohol:

$$\underset{\text{Isocyanate}}{R—N{=}C{=}O} + \underset{\text{Alcohol}}{R'—O—H} \rightarrow \underset{\text{Urethane}}{R—\overset{\displaystyle H}{\overset{|}{N}}—\underset{\displaystyle O—R'}{\underset{|}{C}}{=}O}$$

Polyurethanes are noted for their flexibility, toughness, and abrasion resistance. They are available in two-component systems, moisture curing systems, and air curing systems where drying oil is added to the polymer. The only system of importance in waterborne coatings is the latter. (See Table 5.10.)

Polyvinyl Resins

In a broad structural sense, all resins derived from the vinyl radical (CH_2=

CH_2) may be classified as vinyl resins. Polyvinyl resins are available in three general types:

$$\left[\begin{array}{l} \quad\;\; CH_3 \\ \qquad | \\ O{=}C \\ \qquad | \\ \qquad O \\ \qquad | \\ \;\;-CH-CH_2- \end{array}\right]_n$$

Acetate

$$\left[\begin{array}{l} \qquad\quad\;\; Cl \\ \qquad\qquad | \\ -CH_2-CH- \end{array}\right]_n$$

Chloride

$$\left[\begin{array}{l} \qquad\qquad\quad C_3H_7 \\ \qquad\qquad H\diagup \\ \qquad\qquad\; C \\ \qquad\quad \diagup \quad \diagdown \\ \quad O \qquad\qquad\quad O \\ \quad | \qquad\qquad\qquad | \\ -CH-CH_2-CH-CH_2- \end{array}\right]_n$$

Butyral

Polyvinyl acetate is used extensively in emulsion paints. By itself polyvinyl acetate is a hard resin; the homopolymer must be softened by the addition of 10–20% of a plasticizer such as dibutyl phthalate. To produce a softer resin, vinyl acetate is copolymerized with flexible monomers such as dibutyl maleate, ethylacrylate, ethylene, and so on. (See Table 5.11.)

Polyvinyl acetates have good color retention, and copolymers have good exterior durability. Polyvinyl acetates also have good oil resistance. Many of the other polyvinyl acetate dispersions are coarse suspensions of polymer particles ranging in size from 0.3 to 10.0 microns in diameter; strictly speaking, they are suspension polymers and not emulsion polymers.

Polyvinyl chloride resins by themselves are hard resins and will not form a continuous film. They can be softened with plasticizers or copolymerized with other monomers to produce a softer polymer. Polyvinyl chloride resins have good water resistance and toughness, but have poor light stability. Because of their chlorine content they are generally fire- and flame-resistant.

Polyvinyl butyral resins are exceedingly tough and flexible. Emulsions are produced from the resin by postemulsification.

Vinylidene chloride is utilized in the form of copolymers with vinyl chloride or with acrylonitrile. (See Table 5.12.) It imparts low water absorption and water transmission properties to the copolymer. Because it has a high chlorine content, it has good fire-retardant properties. The polymer is as follows:

$$\left[\begin{array}{l} \qquad\quad\;\; Cl \\ \qquad\qquad | \\ -CH_2-C- \\ \qquad\qquad | \\ \qquad\quad\;\; Cl \end{array}\right]_n$$

TABLE 5.8. POLYAMIDE RESINS—WATER-REDUCIBLE.

Trade name	Producer	NVM % Wt.	Cosolvent	Viscosity, Gardner Holdt	Gardner	Pounds per gal.	Remarks
Genamid[R] 5701	GMCl	65	—	9.13	8	8.6	Polyamide resin for waterborne coatings.
Versamid[R] 5201	GMCl	65	EGMBE	45–65 Pse	9	8.1	Polymide resin for waterborne coatings.

TABLE 5.9. POLYESTER RESINS—WATER-REDUCIBLE.

Trade name	Producer	NVM % Wt.	NVM % Vol.	Cosolvent	Solu-bilizer	Viscosity, Gardner Holdt	Color, Gardner	pH	Pounds per gal.	Acid number solids	Remarks
Cyplex[R] 1600	American Cyanamid	75	69	EGMBE BuOH	—	Z_3–Z_6	8	2	8.9	45	General-purpose spray and coil coatings.
Arolon[R] 465WA8 70	Ashland	70	62	EGMBE	TEA	Y–Z_1	3	8.0	9.0	—	

9200-720-28	Cargill	60	53	—	DMEA	X	5	8.0	8.9	—	High gloss, excellent hardness and mar resistance.
7201-80	Cargill	80	73	—	—	Z_3	5	—	9.1	—	High gloss, excellent hardness and mar resistance.
Coroc[R] RL269-V8	CPV	80	77	EGMBE	—	Z_4–Z_6	11	—	8.63	56	Epoxy ester.
R638M	CPV	50	44	EGMBE	DMEA	W–Y	1	8.0	9.0	55	Industrial baking enamel.
R892M	CPV	45	41	EGMBE	DMEA	Z_4–Z_5	1	8.0	8.8	55	Industrial baking enamel.
Beckosol[R] 13-417	RCI	70	65	EGMBE DGMEE	—	Z_1–Z_3	3	—	9.2	53	Baking enamels.
Kelsol[R] 2080	Spencer Kellogg	75	—	DGMBE IpOH	—	Z–Z_4	3	—	9.0	70	Base resin for E.D. coatings; industrial baking enamels.
Kelsol[R] 3000	Spencer Kellogg	75	—	BuOH	—	Z–Z_4	3	—	8.9	70	High performance electro and conventional spray bake.
DV-3005	Spencer Kellogg	80	—	EGMBE	—	Z_3	2	—	9.2	—	Water-dispersion baking enamel for wide compatability.
DV-3034	Spencer Kellogg	60		IpOH EtOH EGMBE	—	A_2	2	6.7	9.0	—	Baking resin clears with or w/o catalyst.

TABLE 5.10. POLYURETHANE—WATERBORNE.

Trade name	Producer	NVM % Wt.	NVM % Vol.	Viscosity, Gardner Holdt	Pounds per gal.	Notes
Neo Res R-960[R]	PCI[a]	34	31	S	8.8	Aliphatic; resists UV discoloration, abrasion and chemicals
Neo Res R961[R]	PCI	34	31	S	8.8	Aliphatic for roller application on seamless floor.
Neo Res R-962	PCI	34	31	S	8.8	Soft, flexible, aliphatic.

[a]Polyvinyl Chemical Industries.

TABLE 5.11. POLYVINYL AACETATE EMULSIONS.

Trade name	Producer	NVM % Wt.	NVM % Vol.	Viscosity, poises 25°C	Pounds per gal.	Particle size, microns	MFT, °F	Principal co-modifier	Remarks
Amsco Res									
3011	Amsco	55	51	1150	9.0	0.3	39	Acrylic	Trade sales paints
3016	Amsco	55	51	800	9.0	0.3	60	Acrylic	Trade sales paints
3023	Amsco	55	51	900	9.0	0.3	32	Acrylic	Trade sales paints
3077	Amsco	55	51	1500	9.0	0.3	32	Acrylic	Trade sales paints
3077-H-S	Amsco	65	60	2000	9.0	0.3	32	Acrylic	Trade sales paints

Airflex									
500	APCI	55	52	2–5	9.1	0.2	32	Ethylene	Flexible; high binding efficiency; interior, exterior.
510	APCI	55	52	2–5	9.1	0.3	32	Ethylene	Flexible; excellent leveling; interior, exterior
728	APCI	52	48	3–8	9.1	0.2	35	VCl/ethylene	High critical PVC; excellent scrub resistance.
Flexbond									
315	APCI	55	52	8–12	9.1	0.3	38	Acrylate	Good flexibility, excellent water resistance, interior/exterior.
325	APCI	55	52	7–12	9.1	0.3	54	Acrylate	Good flexibility and hardness, interior/exterior.
365	APCI	65	61	10–30	9.2	0.3	—	Acrylate	High solids for use with slurry pigments.
Vinac									
805	APCI	48	44	0.7–2.0	9.1	0.15	5.5	None	Industrial wood intermediate coats.
810	APCI	47	43	0.5–5.0	9.0	0.15	55	Ethylene	Flexible for industrial woods and metal coatings.
885	APCI	48	48	1–3	9.0	0.15	59	None	Similar to 805 but slightly harder.
Polyco[R]									
117-SS	Borden	55	50	11–14	9.2	1.0–3.0	80	None	Interior masonry coatings.
2113	Bordon	55	50	8–16	9.2	0.5–1.5	65	None	Tile coatings—borax-stable.
2151	Borden	55	51	4–10	9.1	0.2–0.8	44	Acrylic	Good leveling interior/exterior.
2151-HS	Borden	65	62	15–25	9.2	0.2–0.6	44	Acrylic	Interior/exterior for use with slurry pigments.

Trade name	Producer	NVM % Wt.	NVM % Vol.	Viscosity, poises 25°C	Pounds per gal.	Particle size, microns	MFT, °F	Principal co-modifier	Remarks
2158	Borden	54	50	5–15	9.0	0.2–0.8	36	Alkyl maleate	Exterior house paints, grain crack resistance.
2370	Borden	54	50	4–8	9.1	0.2–0.6	50	Alkyl maleate	Air dry maintenance coatings.
2381	Borden	45	41	3 max	9.0	0.1–0.2	Bake	Acrylic	High-gloss product finishes, impact, flexibility.
2382	Borden	45	41	1 max	9.1	0.1–0.2	Low bake	Carboxyl	Gloss, low bake hardboard coating.
2392	Borden	55	51	5–10	9.1	0.1–0.5	42	Acrylic	Pigmented basecoats, topcoats, hardboard coatings.
Poly-Tex[R]									
661	Celanese	55	—	7–15	9.1	—	—	Acrylic	Excellent scrub resistance, superior performance.
667	Celanese	65	—	24–40	9.2	0.5–1.5	—	Acrylic	High molecular weight. Excellent scrub resistance, interior/exterior.
6405	Celanese	65	—	7	9.0	—	—	Acrylic	Excellent flow and level scrub resistance.
Daratak[R]									
56L	Grace	55	50	3–6	9.1	1.0	65	None	Very high molecular weight for block-resistant surface coatings.
61L	Grace	55	50	20–40	9.1	1.0	63	None	Borax tolerance for use in fire-retardant coating.

Everflex^{R}									
BG	Grace	51	47	2–5	8.9	1.5	32	Dibutyl maleate	Flexible film for exterior wood and masonry coatings.
G	Grace	52	48	3–7	9.0	1.5	37	Dibutyl maleate	Medium-high flexibility for interior/exterior paints.
GT	Grace	55	51	15–30	9.0	1.0	39	Dibutyl maleate	Fire-retardant coatings.
MF	Grace	52	42	1–5	9.0	0.1	54	Acrylate	Wood and composition board coatings.
E	Grace	55	51	3–7	9.0	0.2	52	Acrylate	General-purpose interior/ exterior paints.
T	Grace	65	61	30–40	9.2	0.25	50	Acrylate	High-quality gloss, semigloss, and flat coatings.
Gelva^{R}									
TS-30	Monsanto	55	50	12–14	9.2	0.5	99	None	High quality interior/exterior primer sealer.
TS-70	Monsanto	55	52	9–13	9.0	< 1.0	50	Maleate	Flexibility in paints.
Wallpol^{R}									
40-133	RCI	55	51	8	9.6	0.3	52	Acrylic	Interior/exterior paints.
40-134	RCI	51	44	6.0	9.6	0.5	70	Acrylic	High gloss and semigloss paints.
40-135	RCI	51	44	6.0	9.6	0.5	75	Acrylic	High gloss and semigloss paints.
40-136	RCI	55	51	10.0	9.6	0.3	49	Acrylic	Interior/exterior paints.
40-140	RCI	65	61	25	9.8	0.5	47	Acrylic	High solids emulsion for interior/exterior paints.
40-141	RCI	51	48	23	9.6	0.4	48	Acrylic	Exterior/house paint, very flexible.
40-142	RCI	55	51	8	9.7	0.3	47	Acrylic	Exterior house paints.
40-311	RCI	58	54	15	9.8	0.8	> 77	None	Interior paints.

Trade name	Producer	NVM % Wt.	NVM % Vol.	Viscosity, poises 25°C	Pounds per gal.	Particle size, microns	MFT, °F	Principal co-modifier	Remarks
Ucar[R]									
130	UCC	58	53	20	9.3	0.8	64	None	Sealers, ceiling tile.
350	UCC	65	—	11	9.2	0.6	46	Acrylic	Interior/exterior paints.
360	UCC	55	51	5	9.1	0.5	57	Acrylic	Interior/exterior paints.
365	UCC	55	51	5	9.1	0.7	54	Acrylic	Interior/exterior paints, high build and good leveling.
366	UCC	55	51	6	9.2	0.4	47	Acrylic	Interior/exterior, excellent coalescence and color acceptance.
560	UCC	55	—	2	9.0	0.7	35	Ethylene	Excellent pigment binder.
5000	UCC	55	—	14	9.1	0.2	47	Acrylic	Intumescent fire-retardant paints.

TABLE 5.12. VINYL CHLORIDE AND VINYLIDENE CHLORIDE POLYMER EMULSIONS.

Trade name	Producer	NVM % Wt.	NVM % Vol.	Viscosity, poises @25°C	Pounds per gal.	Particle size, microns	MFT, °F	pH	Remarks
Polyco 2383	Borden	55	47	1 max	9.7	0.1	—	9.5	Unmodified PVC latex for high-bake industrial topcoats.

2384	Borden	56	50	1 max	9.4	0.1	65	9.5	Plasticized PVC latex for air dry or bake industrial coatings.
2607	Borden	55	51	2 max	9.0	0.3	45	8.0	Acrylic-made PVC copolymer latex for int/ext SG paints.
2610	Borden	56	49	1 max	9.5	0.15	70	9.5	Plasticized PVC latex for factory-applied coating.
2611	Borden	56	48	1 max	9.8	0.15	—	9.5	Unmodified PVC latex for factory-applied coating.
2617	Borden	56	50	1 max	9.4	0.15	70	9.5	Phosphate plasticized PVC latex for fire-retardant coatings.
Saran									
112	Dow	54	43	30 max	10.6	0.13	—	2	Vinylidene chloride copolymer; moisture, vapor, and gas barrier.
118	Dow	53	42	30 max	10.6	0.13	—	2	Vinylidene chloride copolymer; moisture, vapor, and gas barrier.
122	Dow	52	41	30 max	10.6	0.13	—	2	Vinylidene chloride copolymer; moisture, vapor, and gas barrier.
137	Dow	52	41	30 max	10.6	0.13	—	2	Vinylidene chloride copolymer; moisture, vapor, and gas barrier.
143	Dow	54	43	30 max	10.6	0.13	—	2	Vinylidene chloride copolymer; moisture, vapor, and gas barrier.
150	Dow	53	42	30 max	10.6	0.13	—	2	Vinylidene chloride copolymer; moisture, vapor, and gas barrier.
Daran									
220	Grace	61	48	0.4	11.1	0.12	—	3.5	Vinylidine chloride copolymer; moisture, vapor and gas barrier.
225	Grace	61	48	0.2	11.1	0.17	—	3.5	Vinylidine chloride copolymer; moisture, vapor and gas barrier.
226	Grace	61	48	0.3	11.1	0.17	—	3.5	Vinylidine chloride copolymer; moisture, vapor and gas barrier.

Trade name	Producer	NVM % Wt.	NVM % Vol.	Viscosity, poises @25°C	Pounds per gal.	Particle size, microns	MFT, °F	pH	Remarks
229	Grace	61	48	0.3	11.2	0.17	—	2.5	Vinylidine chloride copolymer; moisture, vapor and gas barrier.
305	Grace	56	44	0.3	10.6	0.16	—	2.2	Vinylidine chloride copolymer; moisture, vapor and gas barrier.

TABLE 5.13. STYRENE-BUTADIENE EMULSIONS.

Trade name	Producer	NVM % Wt.	NVM % Vol.	Viscosity, cps	pH	Pounds per gal.	Particle size, microns	Styrene content	MFT, °F	Remarks
Polyco[R] 2430	Borden	48	48	50	9.5	8.5	0.2	67	45	Interior latex paints.
2445	Borden	50	50	100	5.5	8.5	0.2	64	40	Interior specialty coatings.
Dow Latex 308	Dow	48	48	—	10.5	8.5	0.2	64	40	Interior paints; improved binding, low foam.
Darex 515L	Grace	49	48	40	8.0	8.4	0.2	67	35	Highly efficient binder for interior paints and primer sealers.

Polystyrene and Its Copolymers

Styrene monomer is readily polymerized by bulk, solution, suspension, or emulsion methods in the presence of peroxide and heat. Polystyrene is a hard resin whose principal use is in plastics molding. However, styrene is copolymerized with softer materials to produce useful coating materials.

Styrene-butadiene lactices are produced by emulsion polymerization with butadiene acting as a plasticizing monomer.

$$\left[-CH_2-\underset{C_6H_5}{CH}- \right]_n$$

Polystyrene

$$\left[-CH_2-CH=CH-CH_2-\underset{C_6H_5}{CH}-CH_2- \right]_n$$

Styrene-butadiene

Styrene-butadiene latices are low in cost and have good water resistance. (See Table 5.13.)

Styrene monomer is copolymerized with drying oils and alkyds. The styrene speeds the drying of these products.

6

Surfactants

The term "surfactant" is a general term for surface-active agents which includes wetting agents, detergents, emulsifiers, dispersants, foaming agents, penetrating agents and spreaders.

Surfactants have multiple uses in aqueous coatings. First, a surfactant is used as a dispersant during the grinding of the pigment. It provides better wetting of the pigment particles. Second, surfactants are required as emulsifiers either in emulsion polymerization or postemulsification of the vehicle. The emulsifier produces stable interfaces between the oil or other materials and water. The third function of the surfactant is to impart certain physical properties to the aqueous coating such as better wetting to the surface on which it is being applied, better flow, and so on.

A surfactant is a chemical compound that affects the surface forces of a liquid or a solid in relation to other liquids, gases, or solids. The surfactant generally consists of a molecule in which one end is hydrophilic (water-loving) and the other end is lipophilic (oil-loving).

The sodium salts of fatty acids (i.e., ordinary soap) would be an example:

$$\underbrace{CH_3{-}CH_2{-}CH_2{-}CH_2{-}CH_2{-}CH_2{-}CH_2{-}CH_2{-}CH_2{-}CH_2{-}CH_2}_{\textit{Lipophilic}}{-}\underbrace{\overset{\displaystyle O}{\overset{\|}{C}}{-}ONa}_{\textit{Hydrophilic}}$$

When a soap is added to water, it lowers the surface tension (i.e., it reduces the interfacial tension between water and air). The resulting solution foams readily and has other attributes of soap solutions. In addition, the soap promotes wetting of solids, mixing of immiscible liquids, and so forth.

WETTING AGENTS

Wetting agents can be described as substances that lower the interfacial tension between a liquid and a solid surface. Thus, they promote easier wetting of a solid by a liquid. The surface tension of pure water is 72 dynes per centimeter. The use of 0.001 to 0.1% of surfactant may reduce this figure below 30 dynes per centimeter. The surface tension can be determined easily by the DuNuoy tensionmeter.

Wetting agents lower the surface tension of water at the water–air interface.

They may be added to surface coatings to increase spreading and/or penetrating action. Most surfactants lower surface tension and, therefore, have wetting ability under certain conditions. Hence it is wise to specify a product as a "penetrating agent" or a "spreader" or "surface wetter" to define or limit a product more closely. Surface activity is a specific property and will vary from interface to interface. Because a wetting agent may be helpful in causing water to wet a particular hydrophobic material, it does not necessarily follow that it will be effective in promoting speedier wetting of every surface. The Draves Test rates the relative wetting ability of surfactants in respect to textiles but is of little or no value in estimating the ability of any surfactant to wet glass, metal, or other noncellulosic surfaces.

EMULSIFYING AGENTS

An emulsifying agent is a material used to produce more or less stable mixtures of immiscible liquids such as oil and water. It promotes ease of mixing of the oil and water by reducing interfacial tension. Whereas a wetting agent lowers the interfacial tension between a liquid and a solid, an emulsifier lowers it between a liquid and a liquid.

DETERGENTS

A detergent is a material that cleans or aids in cleaning. There are several classes of detergents, not all being surfactants. For example, water-soluble alkalis or alkaline salts having detergent properties, but containing no soap, are called alkaline detergents. Synthetic organic detergents must possess both wetting and emulsifying properties plus some dispersing action (and foaming ability) to aid in removal of oily material and solid dirt from surfaces.

DISPERSING AGENTS

A dispersing agent is used to promote the suspension of fine particles of a solid in a liquid. Hence, both surfactants and colloidal thickeners may fit the definition. A dispersing agent must wet the surface of the solid; yet, at the same time, like an emulsifier, it must form a connecting link between particle and dispersing medium. The molecule is absorbed on the surface of the individual particles leaving exposed molecular ends which all have the same charge. Since like charges repel, the particles separate and remain apart. At a given solids content, such a dispersed system has a lower viscosity than a flocculated system. Since some pigments are hydrophilic and others hydrophobic, no one dispersant will work for all pigments. Good dispersion in waterborne paints provides better hiding, improved leveling, and better color development, and minimizes flooding and floating.

The conditions of use require the same consideration as those of an emulsifying agent, and for this reason it is often hard to draw lines of demarcation. The same agent may be an emulsifier in one case, a dispersant in another, and a wetting agent in a third.

Certain inorganic materials such as phosphate and silicates can act as dispersants for pigments. They are absorbed on the surface of the pigment, producing an electrostatic charge on the particles, which causes the particles to repel each other. The absorption of the dispersant on the surface of the pigment displaces some of the bound water, which results in a very pronounced increase in the fluidity of the slurry or paste. Multivalent ions hinder dispersion. This is particularly true of calcium and magnesium ions found in hard water. For best dispersive action, these interfering ions must be removed from solution. This is also accomplished with the phosphates and silicates.

Polyphosphates:	
Tetrasodium pyrophosphate	$Na_4P_2O_7$
Sodium triphosphate	$Na_5P_3O_{10}$
Sodium tetraphosphate	$Na_6P_4O_{13}$
Sodium metaphosphate	$(NaPO_3)_x$
Silicates:	
Sodium metasilicate	Na_2SiO_5
Sodium disilicate	$Na_2Si_2O_5$

STRUCTURE OF SURFACTANTS

Surfactants are compounds in which one portion of each molecule is hydrophilic (water-loving) and another portion is lipophilic (oil-loving) or hydrophobic (water-fearing). The hydrophilic part of the molecule may be a carboxylate, sulfate, sulfonate, alcohol, or alcohol-ether. The lipophilic portion of the molecule may be a long hydrocarbon chain as in fatty acids, or a straight, branched, or cyclic hydrocarbon of petroleum or an aromatic hydrocarbon containing alkyl side chains.

TYPES OF SURFACTANTS

Surfactants may be classified as anionic, cationic, amphoteric, and nonionic.

Anionic. The anionics bear a negative charge and migrate toward the anode or positive pole while in solution. They are the oldest and best known of surfactants.

$$\left[\mathrm{R{-}\overset{\overset{\displaystyle O}{\|}}{C}{-}O}\right]^{-} \quad \mathrm{M^+}$$

Soap

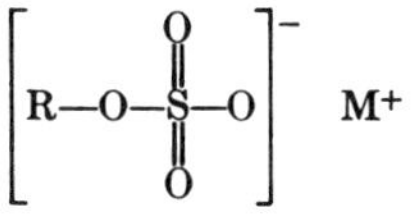

$$\left[R{-}O{-}\overset{\overset{\displaystyle O}{\|}}{\underset{\underset{\displaystyle O}{\|}}{S}}{-}O \right]^{-} M^{+}$$

Alkyl sulfate

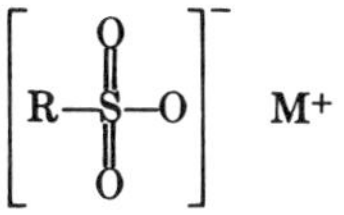

$$\left[R{-}\overset{\overset{\displaystyle O}{\|}}{\underset{\underset{\displaystyle O}{\|}}{S}}{-}O \right]^{-} M^{+}$$

Alkyl sulfonate

Cationic. The cationic group is the opposite of an anionic group. Its surface activity is due to the presence of a long-chain oil-soluble cation. These surfactants are used very little in latex paints, since they would neutralize the charges of an anionic surfactant, which might also be present in the latex, and "break" the emulsion. Cationic surfactants are used to produce stable asphaltic emulsions. Cationic surfactants also possess germicidal, anticorrosive, antistatic properties. The following is the structure of a cationic surfactant:

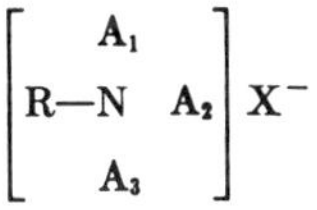

$$\left[R{-}\overset{A_1}{\underset{A_3}{N}}\ A_2 \right] X^{-}$$

R represents a hydrophobic group such as a long-chain aliphatic or aromatic group, X represents a negative ion such as Cl^-, Br^-, I^-, or other monovalent ions, and A_1, A_2, and A_3 represent hydrogen, alkyl/aryl, or heterocylic groups.

Amphoteric. Amphoteric or ampholytic surfactants contain both a positive and a negative charge. These charges may neutralize each other so that at a given pH the surfactant behaves as if it were nonionic. These surfactants usually exhibit cationic properties in acid solutions and anionic properties in alkaline solutions.

Nonionic. Nonionic surfactants depend chiefly upon hydroxyl groups and ether linkages to create the hydrophilic action.

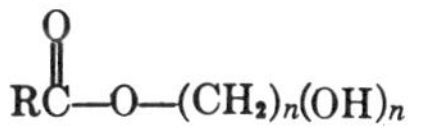

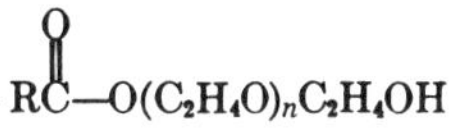

$$R\overset{\overset{\displaystyle O}{\|}}{C}{-}O{-}(CH_2)_n(OH)_n$$

$$R\overset{\overset{\displaystyle O}{\|}}{C}{-}O(C_2H_4O)_nC_2H_4OH$$

These compounds achieve solubility by being hydrated in water solutions. When the temperature is raised, the forces of hydration are reduced, and there is a precipitation of the surfactant, lowering its effectiveness. This shows up as a turbidity or cloud. This may lead to erratic results when processing produces a higher temperature in the paint.

SOLUBILITY AND BALANCE

Two factors are of primary importance in producing surface-active agents to do a specific job: solubility and balance.

The solubility of a surfactant is dependent upon the nature and relative sizes of the hydrophobic and hydrophilic components. An increase in the length of the hydrocarbon (hydrophobic) chain tends to produce greater solubility in oils, while an increase in length or strength of the water-solubilizing group decreases oil solubility.

Balance involves the relation of the hydrophilic group to the hydrophobic portion. If the balance is altered, the solubility is also altered. To illustrate balance, consider a nonionic surfactant made by reacting a fatty acid such as lauric (hydrophobe) with a number of moles of ethylene oxide (hydrophile):

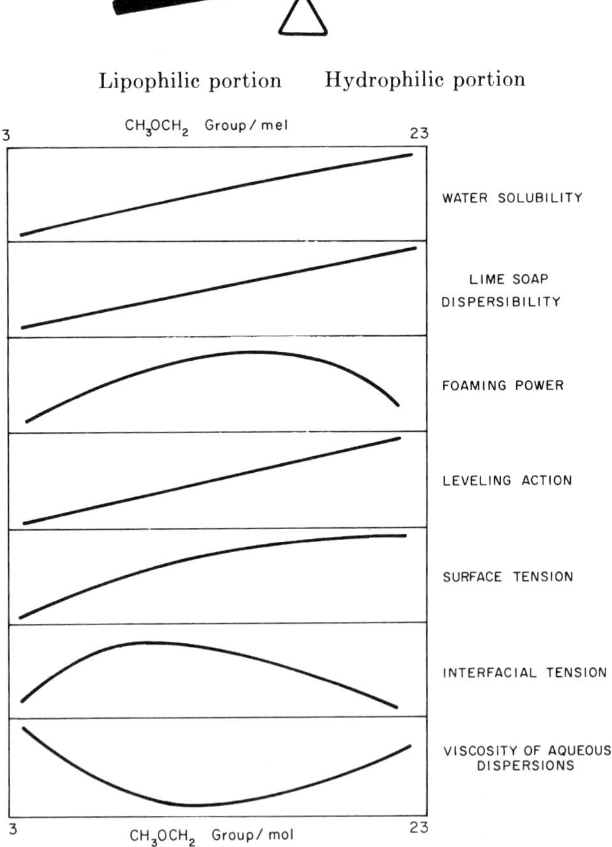

FIGURE 6.1. Surfactant balance.

The effect of the lipophilic hydrocarbon chain on the physical properties of the molecule are shown in Figures 6.1 and 6.2. Surfactants can be illustrated schematically as shown in Figure 6.3.

When, for instance, an anionic molecule is added to a mixture of oil and water (or polar and nonpolar liquids), it orients itself at the interface between the two immiscible liquids. The hydrophilic portion seeks the water phase,

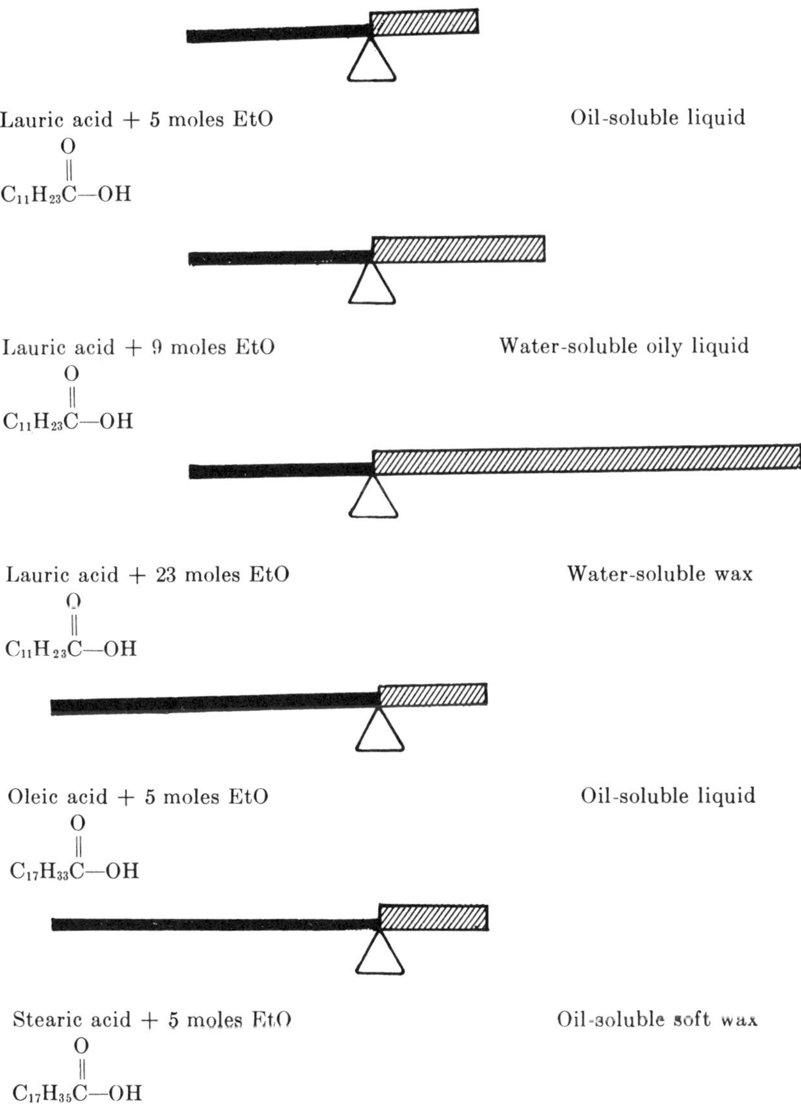

FIGURE 6.2. Effect of balance on physical state of surfactant.

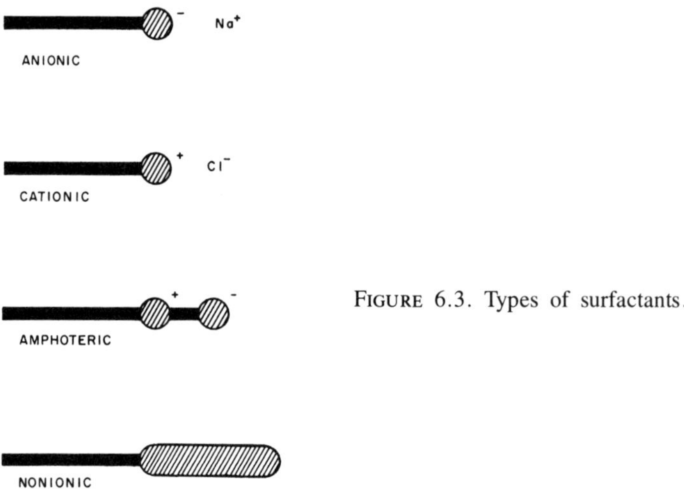

FIGURE 6.3. Types of surfactants.

and the lipophilic seeks the oil phase. (See Figure 6.4.) Figure 6.5 illustrates oil-in-water and water-in-oil emulsions, showing the surfactant orientation for each.

Table 6.1 shows the classification of surfactants. McCutcheon[1] lists almost two thousand commercial types with their trade names, chemical composition, use, type, form, and concentration.

APPLICATION

While surfactants comprise only 0.2 to 3.0% by weight of all aqueous coating, they are an indispensible part of the formulation. They are responsible for aiding pigment dispersion and wetting, color development, emulsification, emulsion particle size, leveling, adhesion, washability, and stability.

The theory of surfactants has been studied and developed to an advanced state as an important part of the field of surface chemistry. The contributions of Adams, Alexander, Harkins, McBain, and others have led to a better understanding of the properties of surfactants. Unfortunately, a large gap exists between our scientific knowledge and our practical knowledge. Since aqueous coatings can contain as many as 15 to 20 different ingredients, interaction between the surfactant and these materials may give unexpected properties. For example, some surfactants raise the viscosity of certain thickeners, while other surfactants lower the viscosity. The pH will have an effect on the performance of a surfactant; and since the pH of aqueous coatings can range from 5.0 for homopolymer PVA to as high as 10.0 for acrylics, this factor must be carefully considered.

As with other organic compounds, the higher the molecular weight is, the

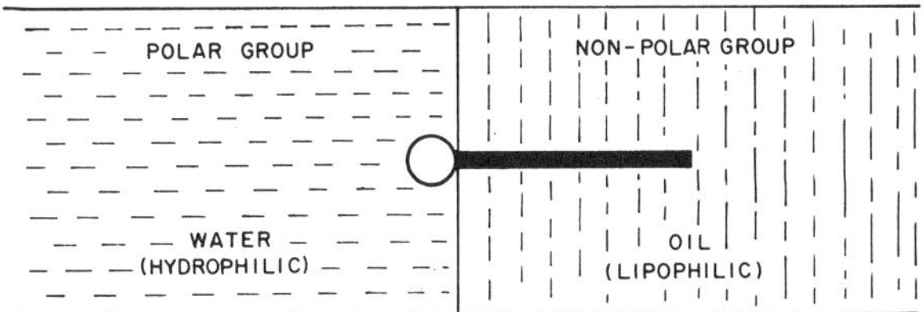

FIGURE 6.4. Emulsion interface.

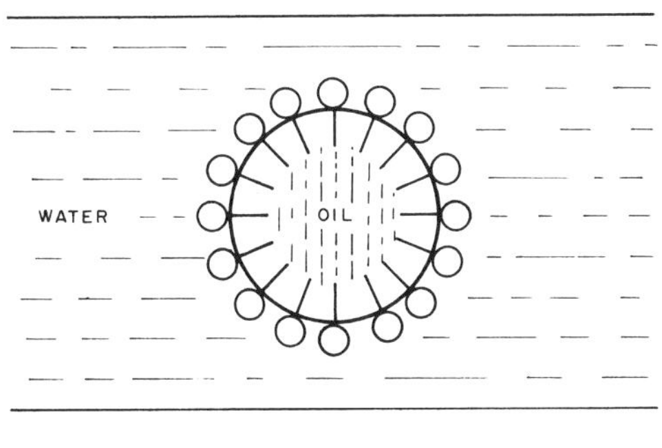

OIL IN WATER O/W

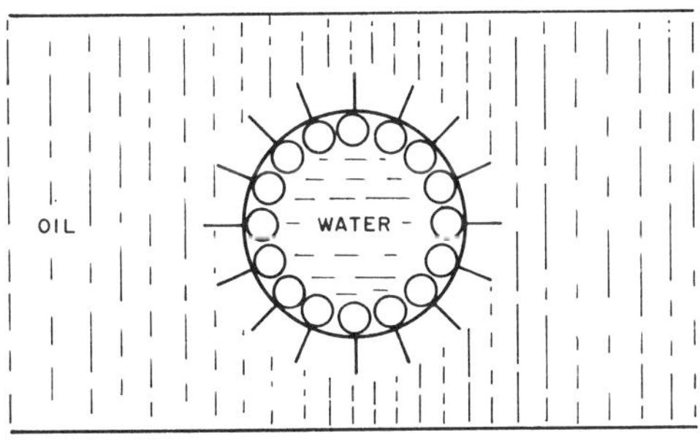

WATER IN OIL W/O

FIGURE 6.5. Emulsions.

less the solubility. Solubility increases with temperature. Nonionic surfactants exhibit a phenomenon known as "cloud point."

Most surfactants lower the surface tension of water, and data are usually available from the manufacturer. Most wetting agents are effective at the 0.2% concentration range. Maximum wetting occurs with a branched-chain molecule. (See Figure 6.6.)

Properly dispersed solid particles remain suspended in a liquid in paints, printing inks, and so forth. The solid particles (i.e. pigment) are first wetted out in the liquid medium, and then the surface of each particle is coated so

TABLE 6.1. CLASSIFICATION OF SURFACTANTS.

- I. Anionics
  - A. Carboxylic Acid
    - (1) Fatty acid, rosin and naphthenic acid soaps
  - B. Sulfuric Acid Esters
    - (1) Alkyl sulfates, alcohols and olefins
    - (2) Sulfated oils and esters
    - (3) Sulfated amides and ethers
  - C. Sulfonic Acids
    - (1) Alkyl sulfonates
    - (2) Alkyl-aryl sulfonates
    - (3) Sulfonated amides and esters
  - D. Miscellaneous—Phosphates, sulfonates, etc.
- II. Cationics
  - A. Simple Amine Salts
  - B. Quaternary Ammonium Salts
  - C. Amino Amides and Amidazolines
- III. Amphoteric
  - A. Amino and Carboxyl Groups
  - B. Amino and Sulfuric Ester or Sulfonic Groups
- IV. Nonionics
  - A. Alkyl, Alkyl-aryl Ethers and Thioethers
  - B. Esters and Amides

that agglomeration does not occur. Protective colloids also play a role in proper dispersion. Anionic or cationic surfactants may be absorbed on solid particles and provide them with a charged surface. Repulsion between like charged particles prevents agglomeration. In the case of sulfonic acid surfactants, dispersing action in aqueous systems increases from alkyl benzene to alkyl naphthalene. It would appear that increased aromaticity of the surfactant structure increases the absorption on solid surfaces. Nonionic surfactants are effective dispersants, particularly the alkyl phenol, ethylene oxide condensates. In this case the hydrophobic groups are absorbed on the solid surfaces, and the polyglycol chains form an envelope around the particles.

Many of the concepts about surfactants as dispersing agents apply equally well to their use as emulsifiers. Hydrophobic groups dissolve in the surface of oil droplets with the hydrophilic groups oriented outward into the water phase. The most important concept in emulsions is that of an interfacial film between the water phase and the oil phase. A strong film prevents coalescense of the dispersed droplets. Charged droplets are stabilized because of repulsion between like charges.

Nonionic surfactants as a class are quite often superior to anionics or cationics for emulsification. The hydrophobic–hydrophilic balance can be easily altered in nonionics by structural changes.

In some cases a special technique is used in making an oil emulsion in which the fatty acid is dissolved in the oil and the amine is dissolved in

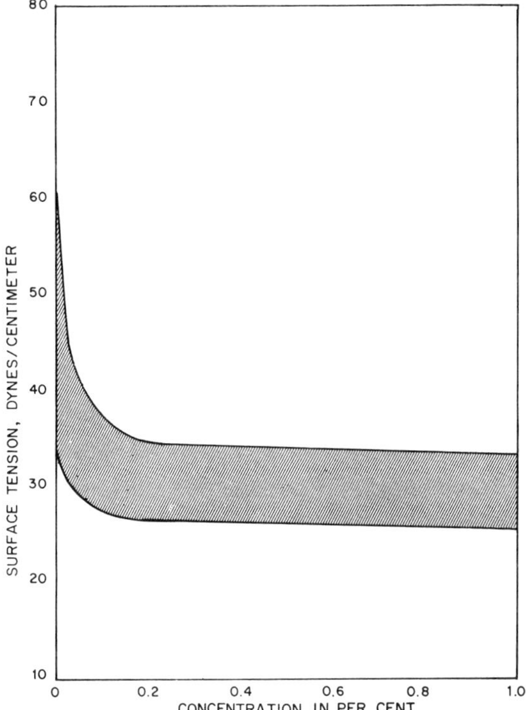

FIGURE 6.6. Range of surface tensions in most surface-active compound-water solutions.

water. A similar technique can be used with nonionics in which an oil-soluble nonionic is added to the oil phase and a water-soluble nonionic added to the water phase. Sometimes a combination of an anionic and nonionic is more effective than a single surfactant.

While foaming is sometimes desirable in cleaning and flotation applications, it is undesirable in coating applications. When this occurs, it is necessary to use a defoamer.

SELECTION OF SURFACTANT

The proper selection of a surfactant can involve a considerable amount of laboratory time. Any method of shortening this time would be a money saver. Probably the most successful approach to this problem was by Griffin[2]. It has been designated the HLB method, which stands for hydrophile–lipophile balance.

Each surfactant has a specific HLB designated by a number. Each material likewise has a required HLB. The problem is to determine these values and fit them together. The HLB number can be related to specific applications as shown in Table 6.2.

The first step in selecting the proper surfactants would be to define as accurately as possible the properties required in the end product. These might be:

1. Type of product (oil-in-water, etc.).
2. Degree of stability required.
3. Permissible cost.
4. Ease of manufacture.
5. Equipment available for use.

The two formulas in Table 6.3 are for calculating HLB of surfactants. The first formula is for ester-type nonionic surfactants. It is based on the saponification number of the ester and the acid number of the fatty acid used to prepare the ester. The term "S over A" is actually a fractional representation of the weight-percentage of the fatty acid (or lipophilic) material in the molecule. The second formula is for nonionic surfactants of either the fatty acid ester or fatty alcohol ether type and is based on the weight percentage of ethylene oxide in the molecule. In some cases it is impossible to calculate HLB from these formulas. In such cases it is necessary to determine the values experimentally in the laboratory. A rough approximation is shown in Figure 6.7. (See also the related Table 6.4.)

In determining the HLB value of the material to be emulsified the first step is to choose two reasonable surface agents, one of low and one of high HLB value. (See Tables 6.5 and 6.6 for some HLB values.) Attempts are then made to prepare a final product with these materials close to the desired prod-

TABLE 6.2. HLB RANGES AND THEIR APPLICATIONS.

Range	Application
3–6	W/O emulsifier
7–9	Wetting agent
8–18	O/W emulsifier
13–15	Detergent
15–18	Solubilizer

TABLE 6.3. FINDING HLB.

1. $$\text{HLB} = 20\left(1 - \frac{S}{A}\right)$$

S = saponification number
A = acid number of separated acid

2. $$\text{HLB} = \frac{E + P}{5}$$

E = weight % ethoxy content
P = weight % polyol content

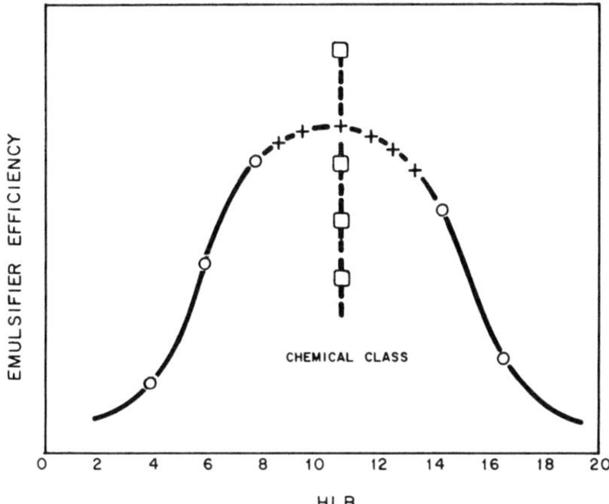

FIGURE 6.7. Determination of HLB of given material.

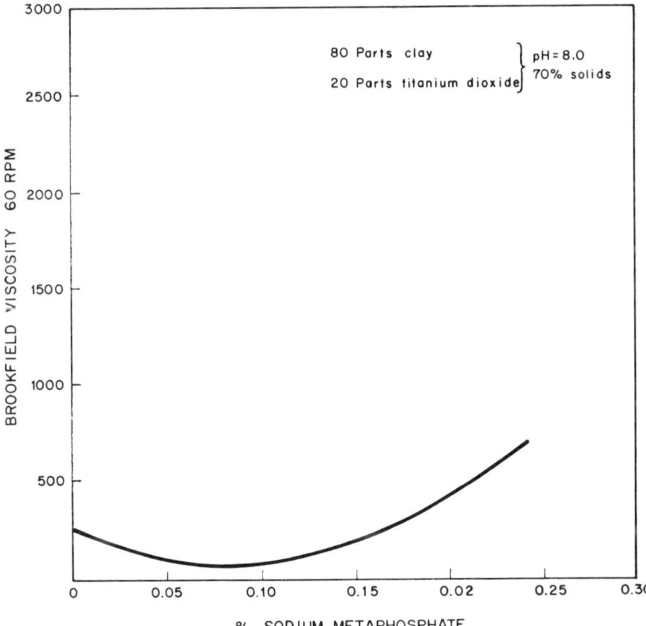

FIGURE 6.8. Viscosity vs. percent dispersant.

uct on equipment that will approximate as nearly as possible plant production methods. Products prepared from either the high or low HLB will probably be unsatisfactory, so blends (usually 80:20, 60:40, 40:60, and 20:80) are made. A point of increased efficiency, based on the standards set for the product, will be noted. At this point blends at narrower increments are run. The point of optimum behavior or maximum efficiency will be recorded as the optimum "required HLB." When the required HLB has been determined, it is logical to test a variety of chemical types or families. Nothing has been said about emulsifier concentration, but usually HLB is not affected by concentration; therefore, if the optimum HLB is determined, the minimum amount of emulsifier required to get stability may be used. In making the usual oil-in-water type of emulsions, the most stable emulsions are usually produced when the initial emulsion is a water-in-oil type that inverts to the desired oil-in-water type during mixing of the ingredients.

When surfactants are used as pigment dispersants, there is an optimum amount that gives the lowest viscosity. This is the amount to use in the formula based on the weight of pigment and extender. (See Figure 6.8.)

TABLE 6.4. SURFACTANT BEHAVIOR IN WATER WITH HLB RANGE.

Action or Behavior when Added to Water	HLB Range
No dispersibility in water	1–4
Poor dispersion	3–6
Milk dispersion after vigorous agitation	6–8
Stable milky dispersion (upper end almost translucent)	8–10
Translucent to clear dispersion	10–13
Clear solution	13+

TABLE 6.5. HLB VALUES OF MATERIALS.

Material	HLB Value
Cottonseed oil	7.5
Carbon tetrachloride	9.0
Paraffin wax	9.0
Mineral oil	10.0
Silicone oil	10.5
Kerosene	12.5
Cetyl alcohol	13.0
Carnauba wax	14.5
Dimethyl phthalate	15.0
Stearic acid	17.0

TABLE 6.6. HLB VALUES OF COMMERCIAL EMULSIFIERS.

Name*	Chemical Designation	Type	HLB
Span 85	Sorbitan trioleate	N	1.8
Span 65	Sorbitan tristearate	N	2.1
Atlas G-1050	Polyethylene sorbital hexastearole	N	2.6
Atlas G-922	Propylene glycol monostearate	N	3.4
Atmul 84	Glycerol monostearate	N	3.8
Span 80	Sorbitol monooleate	N	4.3
Atlas G-2139	Diethylene glycol monooleate	N	4.7
Atlas G-2124	Diethylene glycol monolaurate	N	6.1
Span 40	Sorbitan monopalmitate	N	6.7
Atlas G-2140	Tetraethylene glycol monooleate	N	7.7
Span 20	Sorbitan monolaurate	N	8.6
Tween 61	Polyoxyethylene sorbitan monostearate	N	9.6
Tween 65	Polyoxyethylene sorbitan tristearate	N	10.5
Atlas G-2076	Polyoxyethylene monopalmitate	N	11.6
Atlas G-2127	Polyoxyethylene monolaurate	N	12.8
Tween 21	Polyoxyethylene sorbitan monolaurate	N	13.3
Tween 60	Polyoxyethylene sorbitan monostearate	N	14.9
Atlas G-3820	Polyoxyethylene cetyl alcohol	N	15.7
Atlas G-2129	Polyoxyethylene monolaurate	N	16.3
Tween 20	Polyoxyethylene sorbitan monolaurate	N	16.7
Myrj 53	Polyoxyethylene monostearate	N	17.9

* Altas Powder Company.

REFERENCES

1. "Detergents and Emulsifiers," New York, John W. McCutcheon, Inc., 1973.
2. Griffin, W. C., *Offic. Dig. Federation Paint Varnish Prod. Clubs,* **28,** 446 (1956).
3. Swartz and Perry, "Surface Active Agents and Detergents," Vols. I and II, New York, Interscience Publishers, 1949, 1958.

7
Protective Colloids and Thickeners

Certain water-soluble resins and gums can be used as binders in water-soluble coatings and as protective colloids or thickeners in emulsion paints.

When they are used in emulsion paints, their presence in the water phase aids dispersion of pigment, protects paint against coagulation and settling, and adjusts viscosity.

The water-soluble hydrophilic colloids include materials such as gum arabic, gum tragacanth, starch, sodium alginate, methyl cellulose, hydroxyethyl cellulose, polyvinyl alcohol, ammonia caseinate, sodium polyacrylate, and others. These materials produce high-viscosity solutions at low concentrations. As little as 0.1 to 1% of the protective colloid is usually sufficient.

A portion of the protective colloid incorporated in the grind aids in dispersion. By thickening the mix, high shearing action is obtained, ensuring good mixing. The pigment particles are completely coated with the protective colloid which improves the stability of the paint. A protective colloid is added to an emulsion vehicle or latex in manufacture. The protective colloid, because of the increased viscosity it imparts to the water phase, reduces pigment settling tendencies. This improves stability of the system by minimizing the tendency of the dispersed particles to stick together if subjected to stresses by mechanical means, freezing, and so on. The water-soluble resin, when used as a thickener, produces the correct viscosity for application. The type of protective colloid used will affect the flow and leveling of the coating. For many reasons, including economy, it is advantageous to keep the amount of protective colloid to a minimum. Protective colloids are relatively high in price as compared to other ingredients used in the paint. Once the film is applied and dried, the protective colloid serves no useful purpose and may actually detract from the performance of the coating. Being water-soluble, these materials make the coating more water-sensitive, as shown by scrub tests, humidity tests, exterior exposure tests, and so forth. The protective colloids are usually not compatible with the vehicle or latex, so that when the film dries, there is a tiny network of water-sensitive resin throughout the film.

In some cases, a material such as polyvinyl alcohol will act as an emulsifier as well as a protective colloid, as in the emulsion polymerization of polyvinyl acetate.

The thickener in the external phase also imparts thixotropy characteristics to the paint (see Figure 7.1). Certain thickeners will cause pigment flocculation.

A study made of various thickeners for polyvinyl acetate paints determined that methyl cellulose and hydroxyethyl cellulose possess a combination of overall properties that make them suitable for this use.

Various water-soluble thickeners are selected as flow control additives for various types of emulsion paints. The major thickeners for styrene-butadiene paints are alkali-soluble proteins. They have good flow and leveling characteristics but may cause putrification problems. The cellulosics, especially methyl cellulose and hydroxyethyl cellulose, are the most common thickeners for polyvinyl acetate paints. They have high thickening power for this type of latex. The acrylate salts, casein, and cellulosics are all widely used for acrylic paints.

Latex thickening is not a simple additive function of the latex viscosity plus the thickener viscosity. It is not a simple process of increasing the viscosity of the continuous phase of the dispersion. Rather, it is a combination of several interreactions between the latex and thickener including absorption effects, and so on.

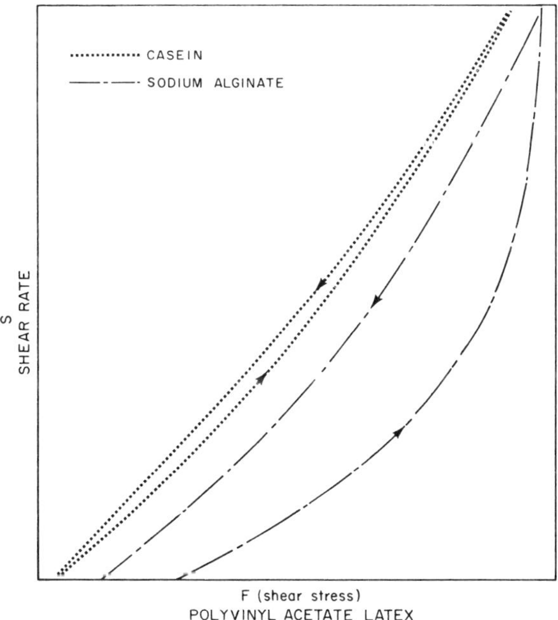

FIGURE 7.1. Viscosity characteristics.

The addition of a thickening agent to a latex or emulsion paint may cause flocculation or coagulation. Some thickeners which are polyelectrolytes may cause coagulation in the same manner as simple salts (see Figure 7.2). Certain thickeners initially produce agglomeration, but as the concentration of thickeners is increased, the agglomerated particles are redispersed.

Protective colloids or thickeners can be classified as shown in Table 7.1. A description of various materials used as protective colloids follows:

Starch. While of little importance in the protective coating field, starch is used to a considerable extent in paper coating. Starch is a mixture of amylose and amylopectin:

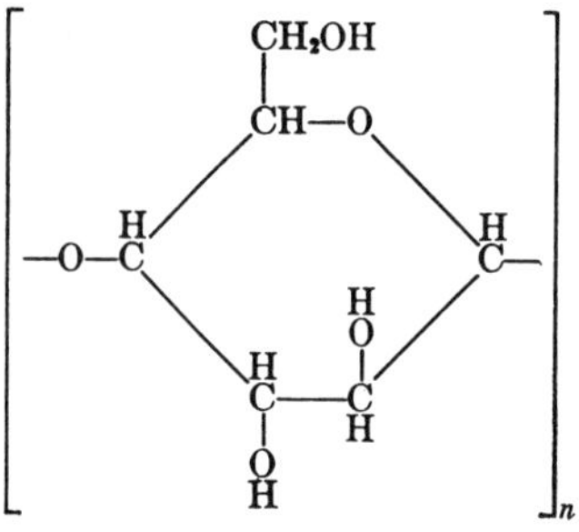

Amylose
20–30% in natural starch

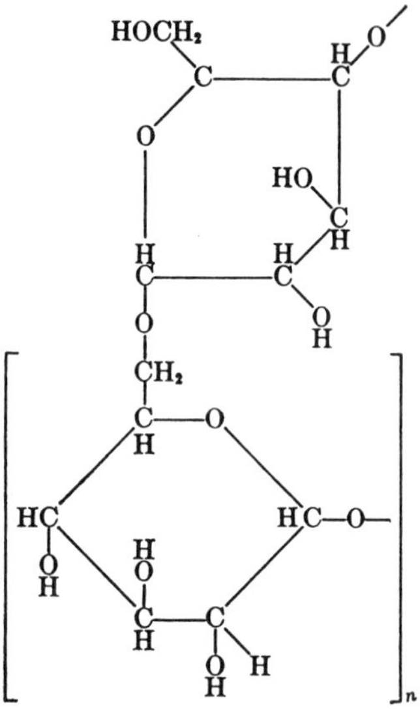

Amylopectin
70–80% in natural starch

Starch is obtained commercially from corn, potatoes, sorghum, wheat, tapioca, sago, and rice.

Starch is found in a unique natural package which is distinctive to the plant that produces it. The molecules of amylose and amylopectin are organized and packed into small granules which vary in size and shape depending on the plant source. The starch can be extracted from plant materials, purified, and modified in an aqueous slurry. On heating in an aqueous slurry, when the temperature reads 140 to 180°F, the starch granules swell and rupture, yielding a viscous colloidal dispersion. Starches are modified in many ways for specific end uses.

Large amounts of starch products are used as surface sizes, binders in clay, and pigment coatings in the paper industry.

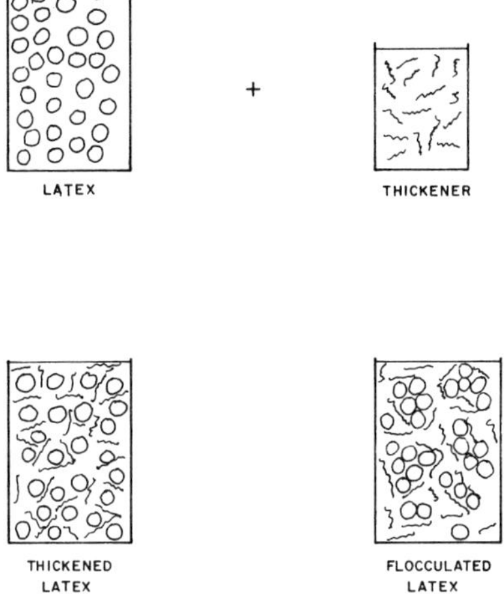

FIGURE 7.2. Mechanism of thickening.

Natural Gums. These include gum arabic, gum tragacanth, gum karaya, and guar gum. These natural resins are usually tree exudations which chemically are high-polymeric saccharides.

Alginates. These are vegetable products obtained from seaweed and kelp. Alginic acid is shown below.

Note that alginic acid differs from cellulose only in having —COOH groups instead of —CH_2OH groups on the polysaccharide chain. While alginic acid itself is insoluble in water, the alkali and ammonia salts are soluble.

TABLE 7.1. COMMON LATEX THICKENERS.

Type	Powder	Solutions	Emulsions
Anionic	Sodium carboxymethyl cellulose Sodium and ammonium alginate Gum-karaya	Sodium and ammonia polyacrylate and copolymers	Polyacrylates
Nonionic	Methyl cellulose Hydroxyethyl cellulose	Hydroxyethyl cellulose	
Amphoteric	Casein Soybean protein Gelatin		

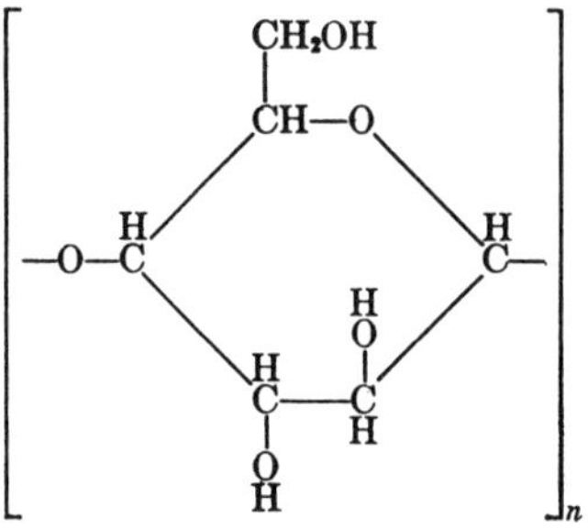

Amylose
20–30% in natural starch

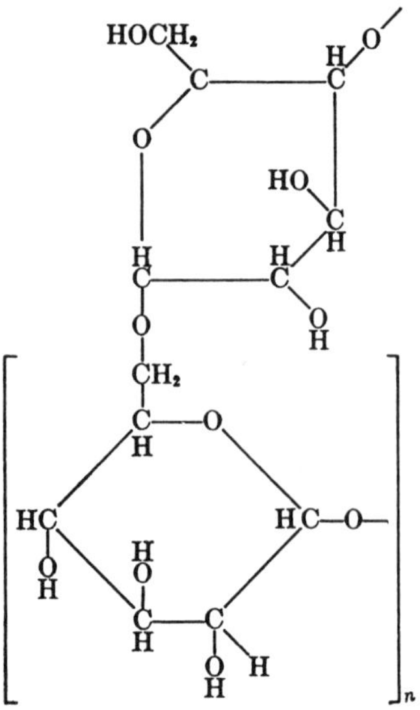

Amylopectin
70–80% in natural starch

Gelatin and Glue. These animal products are the degradation products of collagen, a complex protein material. Both are obtained by cooking bones and skins in water.

Casein. This is the principal protein in milk; for example, cow's milk contains 3% casein. Casein is prepared by heating skim milk to a temperature of 90 to 110°F until it curdles. It is then acidified or treated with an enzyme which precipitates the casein. Casein is a macromolecular substance composed of at least 15 amino acids including glutamic acid, hydroxy glutamic acid, phenylalanine, histidine, alanine, trytophan, proline, and methionine.

As a protective colloid in paint casein gives good pigment dispersion and leveling. In certain formulations, it is possible to produce a higher gloss by the use of casein. Casein is lower in viscosity than most protective colloids so that greater amounts are usually used.

Alpha and Gamma Protein. These types of proteins are isolated from soybeans. They fall in the same general family of chemicals as casein but differ in that milk casein is a complex protein containing, among other things, a phosphorus group, whereas alpha protein would be classified as a simple protein consisting largely of glycinin. Solutions or dispersions are prepared by heating protein with alkali to a temperature of about 180°F; 10 to 12% dispersions are prepared using 6 to 7½% sodium hydroxide based on weight of dry protein.

A preferred pH range for optimum dispersions is 10.5 to 12.5. When the protein is dispersed, the pH of the dispersion is generally adjusted to 8 to 10 by means of buffering acids such as boric acid. Protein is lower in cost than casein. Some emulsion paints containing protein as the protective colloid will lose viscosity on aging; the inclusion of aldehydes will stabilize viscosity.

Cellulosics. Cellulose itself is water-insoluble, but by various chemical treatments it can be made water-soluble. The cellulose molecule is as follows:

```
[          H                ]
           O
      H    |
      C————C
     /|    H\
    / O      \
   /  H       \
—CH            C—O—
   \           H
    \H        /
     C————O  /
     |
     CH2OH                  ]n
```

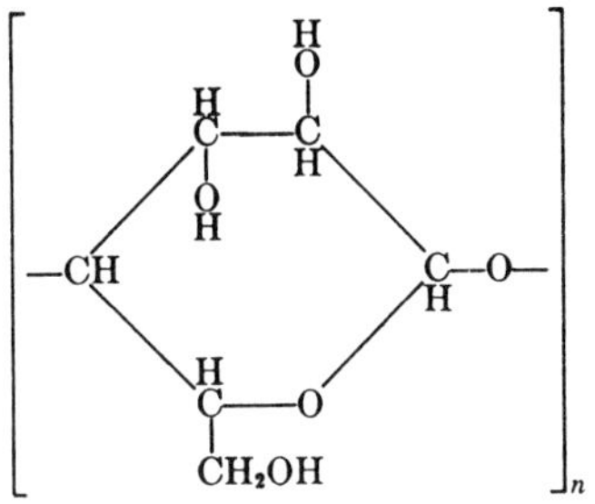

Some of the water-soluble derivatives of cellulose are methyl cellulose, hydroxyethyl cellulose, carboxymethyl cellulose, and so on.

Methyl cellulose is prepared as follows:

R—OH	+	NaOH	→	R—ONa	+	H_2O
Cellulose		Sodium hydroxide		Alkali cellulose		Water
R—ONa	+	CH_3Cl	→	R—OCH_3	+	NaCl
Alkali cellulose		Alkyl halide		Methyl cellulose		

The methyl cellulose molecule is as follows:

```
[       CH2OCH3             ]
        |
        C————O
       /H     \
      /        \
    H/          \
   —C            C—O—
     \          /
      \H   H   /
       O   H  /
       C———C
       H   |
           OCH3             ]n
```

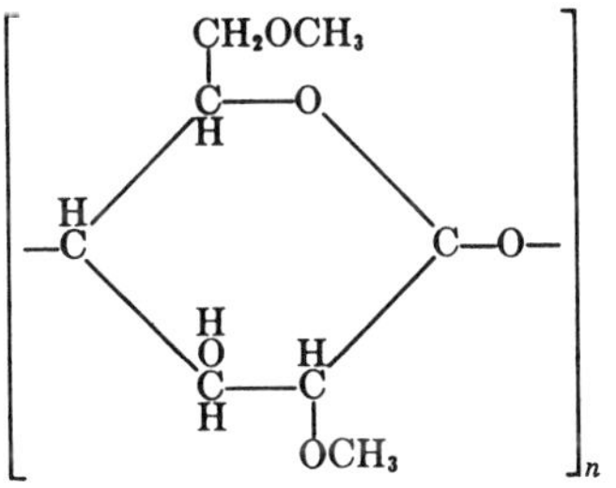

The effect of substitution is to disorder and spread apart the cellulose chains so that water or other solvents may enter to solvate the chain. By controlling the amount and type of substitution, it is possible to produce a range of widely varying properties. With methoxyl substitution, varying the methoxyl content yields a series of methyl cellulose ranging from alkali-soluble to organic-soluble materials. The water-soluble commercial derivatives contain approximately two methoxy groups per anhydroglucose ring.

Methyl cellulose is available commercially in ranges from 10 to 15,000 cps (viscosity cps of 2% solution in water at 20°C). Methyl cellulose is unusual gum in that, as it is heated, its viscosity increases until it gels.

Since the methyl celluloses are insoluble in hot water, an ideal method of dissolving is to disperse the methyl cellulose powder in hot water (190°F) and then add cold water to dissolve it. Charring temperature is 295 to 305°C.

In general, 1 to 2% of a 4000 cp grade methyl cellulose based on paint solids will give good pigment dispersion and stable viscosity and suspension properties to the paint. (See Figure 7.3 for viscosity concentration chart.)

Hydroxyethyl cellulose is also produced from alkali cellulose, by reaction with ethylene oxide as follows:

R—ONa	+	Ch_2OCH_2	→	$ROCH_2CH_2OH$
Alkali cellulose		Alkylene oxide		Hydroxyethyl cellulose

Hydroxyethyl cellulose is available commercially in grades from 10 cps to 4400 cps (viscosity of 2% solutions at 20°C in water). The viscosity of hydroxyethyl cellulose solutions decreases with temperature. Charring temperature is 250°C. The compatibility of hydroxyethyl cellulose with inorganic salts is better than that of other commercial hydrocolloids. In general, 1 to 2% of hydroxyethyl cellulose based on paint solids is used in a paint.

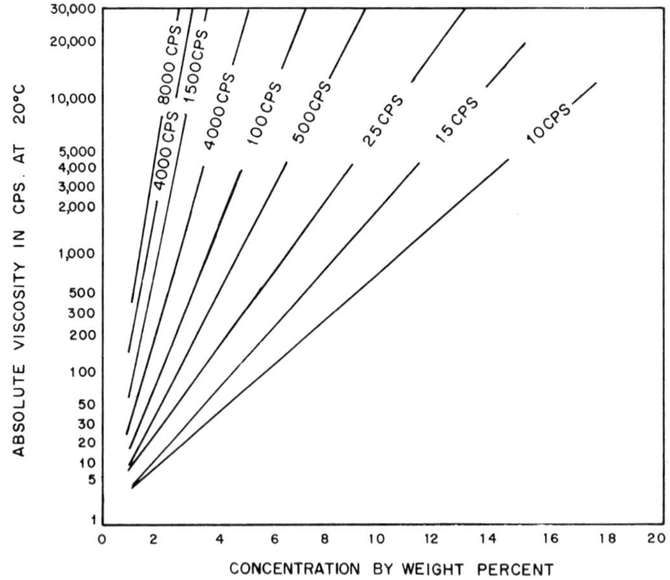

FIGURE 7.3. Viscosity concentration chart for all types of methyl cellulose.

Carboxymethyl cellulose is sold as the sodium salt so that chemically it is sodium carboxymethyl cellulose. It is prepared from alkali cellulose as follows:

$$\underset{\text{Alkali cellulose}}{R\text{—}ONa} + \underset{\text{Sodium monochloroacetate}}{ClCh_2COONa} \rightarrow \underset{\text{Carboxymethyl cellulose}}{ROCH_2COONa} + NaCl$$

There are three hydroxyl groups available for substitution on the cellulose molecule, and commercial water-soluble types of CMC are substituted in the range of 0.5 to 1.5. Carboxymethyl cellulose solutions show a decrease in viscosity with an increase in temperature. Carboxymethyl cellulose is available in viscosity ranges of 10 to 10,000 cps (viscosity of 2% solution in water at 20°C). About 1 to 2% of high-viscosity CMC based on paint solids is used in emulsion paints.

Poly(acrylic) Acid, etc. Included in this class of materials are poly(acrylic) acid and poly(methacrylic) acid; there are salts and copolymers of acrylic acid and methacrylic acid and other monomers. The structural representation of the acrylic polymer is shown below:

$$\text{-------}CH_2\text{-------}\overset{COOR'}{CR}\text{------------}CH_2\text{---------}\overset{COOR'}{CR}\text{---------}$$

where R is H or CH_3 and R′ is H, CH_3, C_2H_5, Na, K, NH_3, etc. Acrylic-type polymers are the most effective thickening agents available in the concentration range up to 0.5%. They are ideal dispersants for pigments, particularly of the inorganic type. Water-soluble acrylic polymers are very effective thickening agents for latex-type systems. With any given water-soluble polymer, thickening of the latex is increased with increased molecular weight of the thickener, which makes the phenomenon simply one of thickening the aqueous phase. However, very low-molecular-weight polymers that have relatively small thickening effect in water may be extremely effective in thickening certain latexes. The variation in thickening action may be attributed to the various latex compositions, particularly the nature of the emulsifier. This phenomenon may become even more complex in a paint where the thickener may react with pigments, inerts, and so forth.

Acrylic polymers are also available in emulsion form for use as thickeners. They are acid-containing cross-linked emulsion copolymers at low viscosity so that they are easy to handle. Solids are 20% or more with a pH of 3.5. Upon the addition of alkali the emulsion converts to a solution of polyacrylate salts which are effective thickeners.

SUMMARY

The choice of thickener affects the liquid properties of an emulsion paint as well as the final properties of the coating. The proper selection of a thickening agent is needed to get the correct rheological properties such as good brushing, good leveling, minimum of sag, and elimination of roller spatter. Some of the cellulosic thickeners are difficult to dissolve and are subject in the package to breakdown from enzymes. Coatings containing thickeners with permanent alkalis such as sodium are more susceptible to moisture absorption.

8

Pigments and Extender Pigments

Pigments are used in coatings to provide color, opacity, and sheen. They also affect the viscosity, flow, holdout, toughness, durability, and other physical properties of the coating.

Pigments can be divided into two broad classes: colored pigments, such as red, yellow, blue, etc., and white pigments. Colored pigments absorb a portion of the light that falls upon them and reflect back to the eye the remaining color bands which determine the color. Black pigments absorb all of the light. White pigments, while they absorb some light and convert it to heat, do not select particular light bands and, thus, give an appearance of light.

OPACITY

Opacity or hiding power is the ability of the pigment to reflect light. With colored pigments, it is obvious that the hiding power of the pigment is due to its ability to absorb certain light rays as heat and to reflect others. With a white pigment, the light enters the pigment particle.

The hiding power of a white pigment is dependent on two factors: refractive index and particle size. Refractive index is actually a ratio that expresses the relative speed of light in that substance as compared to the speed of light in air. Light rays passing from one medium to another of a different refractive index are bent or "refracted" at the interface; the greater the difference in the refractive index, the more the light rays are bent. The refraction of light by a white pigment takes place when light passes from the paint vehicle into the pigment particle and again when light leaves the crystal after having been transmitted through it. As a result of the many multiple refractions, the light is finally returned to the direction from which it came, or "reflected."

The relationship between the percentage of incident light that is reflected by a given pigment vehicle system and the difference that exists between the refractive indices of pigment and vehicle is expressed in its simplest form by this equation:

$$\text{Reflection coefficient} = \frac{(n_1 - n_2)^2}{(n_1 + n_2)^2}$$

where n_1 = pigment refractive index and n_2 = vehicle refractive index.

An extender pigment, such as silica, which has a refractive index of 1.55, has definite hiding power when surrounded by air, which has a refractive index of 1.0. Therefore, when silica is used in a paint film that has a deficiency of binder, silica has some hiding properties. However, when used in a paint with sufficient binder, which might have a refractive index of 1.50, the hiding power of the silica is almost nil.

High hiding pigments such as titanium dioxide have a much higher refractive index. For example, rutile dioxide has a refractive index of 2.76.

Particle Size. Particle size is important in relation to hiding because the finer the particle size, the more interfaces present for refraction per unit volume. (See Figure 8.1.) There is, however, a limit to the process of reducing particle size to increase opacity. To reflect light of any given wavelength efficiently, a particle must possess a diameter approximately one-half of that specific wavelength. Since a white pigment must be able to reflect all wavelengths of visible light, the particle size range should be 0.2 to 0.4 microns (one-half of the 4000 to 7000 angstrom unit range of visible light). If a particle is made any smaller, a wave of light can simply go around it without being refracted.

Hiding Power. Hiding power is usually expressed in square feet of coverage per pound of pigment. The tinting strength of a white pigment is an entirely empirical figure derived from the amount of a standard blue or black tinting material with which the pigment tested must be mixed to match the depth of blue or gray color obtained with an arbitrarily selected white standard.

One method for measuring the hiding power of paint is the "contrast ratio"

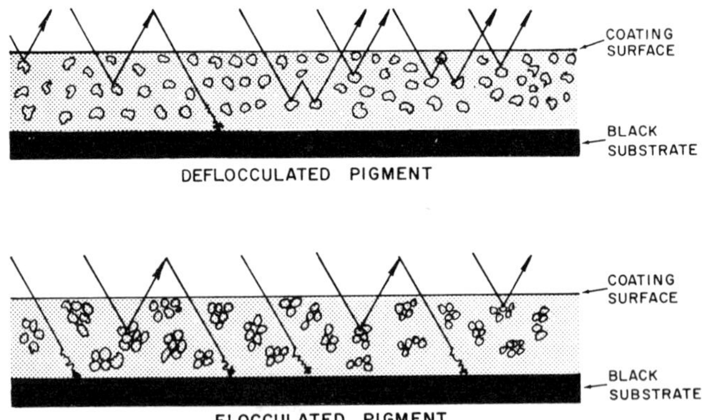

FIGURE 8.1. Hiding Power of deflocculated vs. flocculated pigment.

method. A weighed amount of paint is applied to the one-square-foot surface of a Morest Hiding Power Chart which is divided into black and white squares that have a reflectance of no greater than 5% over the black and 80% over the white. Charts are coated with increasing amounts of paint. The film thickness may be calculated from the weight of paint applied and the specific gravity of the paint solids. Reflectance measurements are made on the dried coatings over the black and white squares. When the ratio of the reflectance over the black to the white equals 0.98, the coating is considered to have complete hiding.

In a study of the hiding powers of ten white pigments in an acrylic vehicle, measurements were made at seven different concentrations. Kubelka-Munk scattering and absorption coefficients were determined. Coarse particle size of the pigment favors the development of a high scattering coefficient at high concentrations. Scattering coefficients increased in a linear manner with concentration up to a particular limiting concentration for each pigment. This limiting concentration is that at which each pigment has the same edge-to-edge separation between particles. The absorption coefficients increased in a linear manner with concentration.

PARTICLE SHAPE

Pigment particles are of various shapes. Their form affects the performance of the coating in both the wet and dry states.

Five general groups as to particle form are spheroidal, lamellar, nodular, acicular, and cubical.

Spheroidal form is found in carbon blacks. Lamellar or flake-type materials are metallic aluminum and micas, which tend to overlap in the coating giving a tight impermeable film. Nodular or rounded irregular shapes are produced by the breakdown in grinding of layer crystals, an example being calcium carbonate. Acicular pigments are needle- or rod-shaped, and an example of this type is zinc oxide. Cubical-shaped pigments are in their original crystal form.

The shape of the pigment affects the viscosity of the wet paint. In general, lumellar and acicular forms produce greater thixotropy than do nodular or spherical forms.

OIL AND WATER ABSORPTION

When a pigment is dispersed in a vehicle, the surface of the pigment particles is wetted with the vehicle, and the voids are filled with vehicle. The oil absorption value of a pigment is determined by adding linseed oil dropwise to a weighed amount of pigment until the mixture forms a pasty mass on mixing with a spatula. Since pigments vary as to the amount of oil or vehicle they require to reach the same state of flow, the oil absorption value is an indication of the viscosity obtained with different pigments. Water absorption of a pigment is determined in the same manner as an oil absorption test except that water is used instead of linseed oil. Oil and water absorption values are not comparable as shown in Table 8.1.

INCORPORATION INTO PAINT

Pigments for water-base paints can be obtained in the form of dry color or in some cases as press cake, pulp color slurries, or pigment dispersions.

Organic pigments are generally precipitated or formed in an aqueous system, and the pigment is filtered off to remove as much water as possible. The press cake has a pigment content of 20 to 30%. The press cake is stiff, and the pigment is not well dispersed and should be ground for use in coatings.

These press cakes may be ground with a dispersant and additional water to produce a pigment dispersion which can be stirred into a paint batch. Pigment content is 12 to 20%. It is a difficult problem to prepare pigment dispersions that are compatible with all polyvinyl acetates, styrene-butadiene, acrylics, and so forth.

Dry pigments that are ground in are the most economical. To determine if a pigment in a colored paint is properly dispersed, the paint is brushed out on a panel, and, as it is drying, the surface is rubbed gently with the index finger. If there is a color difference between the rubbed and unrubbed sections, poor color dispersion is indicated. Regrinding the batch is the only solution to the problem. Paints with poor pigment dispersion will show poor performance, particularly on exposure. (See Figure 8.2.)

WHITE PIGMENTS

White pigments are divided into two classes, the primary or hiding pigments having refractive indexes substantially above 1.50 (the refractive index of

TABLE 8.1. PIGMENTS COMPARISON OF OIL AND WATER ABSORPTION.

	Oil Absorption	Water Absorption
Rutile—non-chalking TO_2	22.0	37.0
Red iron oxide—synthetic	24.0	37.5
Silica—amorphous	30.0	25.5
AA Mica—dry ground	43.0	34.5
Hydrite flat clay	32.0	33.0
Talc #21	34.0	30.0
Ca carbonate—wet ground	20.0	21.0

Grams per 100 g pigment.

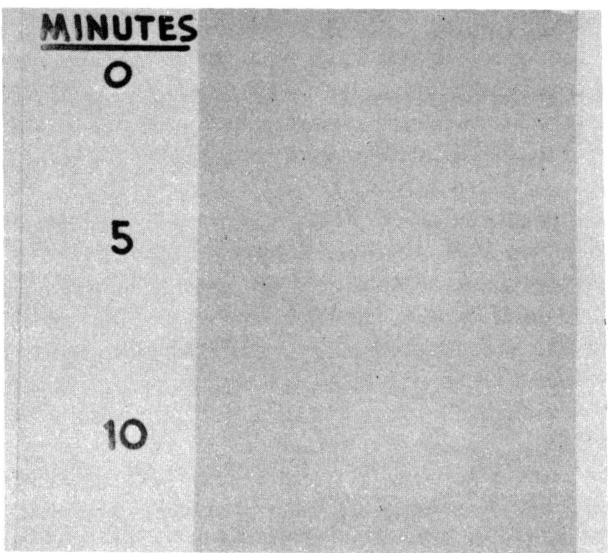

FIGURE 8.2. Pigment dispersion: paint brushed out and rubbed with finger. Note difference in color after five minutes, showing poor dispersion.

most paint vehicles). These are titanium dioxide, zinc oxide, zinc sulfide, antimony oxide, basic carbonate white lead, and basic sulfate white lead. The second class of white pigments, known as extender pigments, has refractive indexes close to that of common paint vehicles (i.e., around 1.50). Pigments of this type are barium sulfate, calcium carbonate, silica dioxide, magnesium silicate, and aluminum silicate.

The type of pigment and the amount of pigment determine the sheen of a paint. A gloss paint contains enough binder so that all pigment particles are completely encased in vehicle, giving a smooth glossy surface. In a flat paint, particles of the pigment which are not covered by vehicle protrude at the surface, breaking up reflected light and producing a dull surface. (See Figure 8.3.)

Pigments act as ultraviolet light absorbers, thus increasing the life of the coating. The inclusion of fibrous or micaceous magnesium silicates in exterior

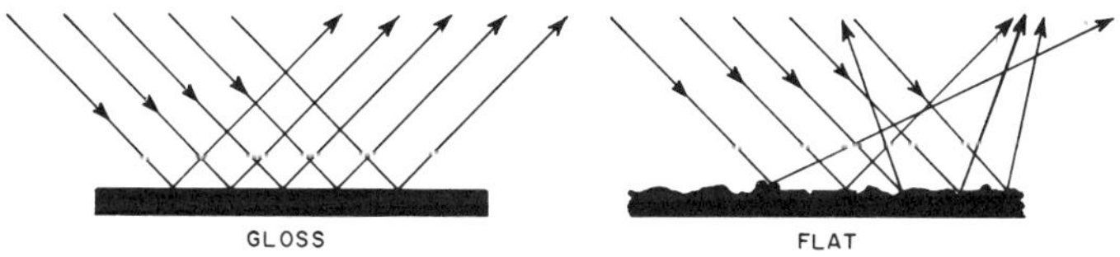

FIGURE 8.3. Gloss phenomena.

house paints adds nothing to the hiding power but definitely improves the durability. The use of corrosion-inhibiting pigments, such as zinc chromate in a coating, develops some passivity in the metal, thus reducing the amount of corrosion. Pigments increase the viscosity or consistency of paints and thus give better application properties, allowing an adequate film to be applied without sags.

Economics is, of course, a factor in the selection of the type and amount of pigment. When enough hiding pigment is present to get satisfactory hiding at the desired film thickness, the remainder of the bulk can usually be made up with extender pigments. (See Table 8.2.)

White Hiding Pigments. The white hiding pigments are inorganic compounds of titanium, zinc, lead, and antimony. The principal use of antimony oxide is in fire-retardant paints. In some instances, these pigments can be deposited on inert materials to produce the so-called let-down pigments, such as lead silicate and titanium calciums. Properties of white pigments are shown in Table 8.3.

Titanium dioxide pigments are produced in two different crystal forms: rutile and anatase. The rutile form has greater hiding and chalk resistance. Titanium dioxide is obtained chiefly from a black ore known as ilmenite, which contains titanium dioxide and iron oxide, plus impurities. In the sulfate process, the ore is treated with sulfuric acid, which separates the iron out as ferrous sulfate. The titanium solutions are then hydrolyzed, purified, and calcined to obtain titanium dioxide. The second method of producing titanium dioxide, which was announced in 1958, is the chloride process. By this process, the ore is treated with chlorine gas to produce titanium tetrachloride, which is a liquid. This is then oxidized in the vapor phase to produce titanium dioxide and gaseous chlorine.

The surface of the titanium pigments can be treated with materials such as zinc oxide, aluminum oxide, antimony oxide, and silicon oxide to make it more usable in paint. The pigment surface can be modified to make it more hydrophobic or hydrophilic to facilitate dispersion in organic liquids and water respectively.

Titanium-calcium pigment is produced by precipitating titanium dioxide on the surface of calcium sulfate particles. It is available at 30 and 50% titanium

TABLE 8.2. INORGANIC WHITE PIGMENTS

White	White—Extenders
Titanium dioxide	Barytes
White lead	Whiting
Zinc oxide	Talc
Lithophone	Mica
	Clay
	Diatomaceous silica

Table 8.3. Properties of White Pigments.

White Pigment	Tinting Strength	Hiding Power	Refractive Index	Gal/lb	Oil Absorption
Rutile titanium dioxide	1600	147	2.76	2.86	18–22
Anatase titanium dioxide	1250	115	2.55	3.08	20–25
50% rutile titanium dioxide–calcium sulphate	880	82	—	3.45	20
Zinc sulfide	640	58	2.37	3.00	22
30% rutile titanium dioxide–calcium sulphate	600	57	—	3.69	22
Lithophone	280	27	—	2.79	12–18
Antimony dioxide	300	22	—	2.13	11–12
Zinc oxide	210	20	2.08	2.14	12–25
35% leaded zinc oxide	175	20	—	2.06	12
Basic carbonate white lead	160	18	2.00	1.78	8–15
Basic sulfate white lead	120	14	1.98	1.90	10–15
Basic silicate white lead	90	12	—	3.00	15

dioxide contents. Titanium-calcium pigments are not generally used in water-type paints because calcium sulfate is partially water-soluble, and the calcium ions present can react with other materials present in the paint, such as protein, and so on. Londergan reported on the use of titanium dioxide in PVA paints.

Zinc oxide pigments are produced by fume processes. The French process uses zinc metal, and the American process uses zinc ore as the source of zinc. Zinc oxide has some fungicidal properties when used in coatings. A vinyl acrylic copolymer latex is now on the market which can be used with reactive pigments such as zinc oxide and leaded zinc oxide. Glycolamine is reported to stabilize zinc-oxide-latex coating compositions. Other stable water-base coating compositions containing zinc are reported. Zinc oxide and titanium dioxide can react in a water system to cause a rapid increase in viscosity.

Lithophone is a composite pigment of zinc sulfide coprecipitated with barium or calcium sulfate.

Zinc sulfide has the greatest hiding power of zinc pigments but is not competitive with titanium pigments on a cost basis for equal hiding. The use of zinc sulfide pigments in aqueous dispersions is reported.

Basic carbonate of white lead is used principally in exterior oil paints and has very little application in water-base paints.

Basic sulfate white lead is slightly lower in cost than basic carbonate of white lead but is also lower in opacity. It is generally used in the same places as basic carbonate of white lead.

Basic silicate white lead is the newest of the lead pigments. It consists of a core of silica surrounded by a surface layer of monobasic lead silicate and monobasic lead sulfate. This produces a low-cost reactive white pigment. It is used to some extent to reduce or prevent "cedar staining" in water-base paints.

Extender Pigments. There are many low-cost extender pigments used in coatings. These are tabulated in Table 8.4. Extender pigment effects are largely governed by the average particle size. The selection of extender affects the color uniformity, gloss, enamel holdout, polishing, and soil removal of interior paints. The proper selection of extender pigments is necessary to attain maximum durability in an exterior paint. Fine-particle-size extenders can sometimes be used to replace a portion of the titanium dioxide in a paint with no loss in hiding. The use of more than one extender in an emulsion paint gives improved performance. Some of the common extender pigments are the following:

Talc is produced by grinding and processing natural deposits of talc. The shape of particles is both fibrous and micaceous. It has widespread use in paints because of its chemical inertness, low cost, good settling characteristics, and good durability. It is not normally used in high-gloss paints.

Calcium carbonate is obtained from natural deposits and from chemical reactions. Calcium carbonates are widely used because of high brightness, low specific gravity, and low cost.

Barytes is made by grinding natural deposits of barytes. Chemical composition is barium sulfate. It is used where its chemical inertness and high specific gravity are advantageous. Because of its high specific gravity, it may show settling tendencies in certain formulations.

Clays obtained from natural deposits are used in water-type paints because of their thixotropy properties and dry hiding power.

Mica is obtained by grinding and processing natural deposits. Mica improves crack resistance and brushing characteristics of a paint. Reports detail its use in emulsion paints.

TABLE 8.4. EXTENDER PIGMENTS.

Name	Composition	Bulking G/100 F	Oil Absorption	Particle Size	Form	Refractive Index
Talc	(magnesium silicate)	4.21–4.44	32–50	1–12	fibrous-flaky	1.59
Whiting	(calcium carbonate)	4.43–4.52	15–25	.05–8	granular	1.63
Barytes	(barium sulfate)	2.70–2.80	8–10	8–10	granular	1.64
Clays	(hydrated aluminum silicate)	4.35–4.65	35–50	1–7	plate	1.56
Mica	(potassium aluminum silicate)	4.20–4.30	28–53	(325 mesh)	lamellar	1.59
Silica	(silicon dioxide)	4.48–4.54	22–30	6–7	granular	1.55
Wollastonite	(calcium silicate)	4.10–4.15	26	7–9	acicular	1.57
Lucite	(diatomaceous silica & whiting)	4.70–4.75	38–46	5–10	granular	1.59

Diatomaceous silica is made by grinding and processing deposits of diatoms, geological remains of unicellular plants. Its high binder demand produces paints of low sheen. It has good chemical inertness.

Extender Pigments in Water Systems. Clays may cause thixotropy, colloidal grades of silica may cause gelling, and the long-term stability of calcium carbonate is questionable. Silicates are preferred for best results, barytes is best for salt spray, and mica tends to reduce film checking and cracking.

SLURRIES

Titanium dioxide and some inerts are now available in water slurry form. The slurry is delivered to the plant in tank cars or tank trucks, pumped to storage tanks and then to the paint-production equipment.

Advantages

The advantages of slurries are that production is speeded up, and labor costs are reduced. (See Table 8.5.)

Slurries can be pumped into the batch much faster than by adding 50-pound bags to the grinding tank. Additional time is saved by adding the slurry as part of the letdown, thereby eliminating the dispersion step.

Titanium dioxide in the form of slurries is about 1¢ per pound cheaper than in the dry form. Handling costs are lower because there are no bags, pallets, and forklift trucks. Power requirements for the dispersion operation are eliminated. Warehouse space for bags is eliminated. Ecology is improved because there is no dust going into the atmosphere from dry operations.

An investment of $30,000 to $200,000 is required for a slurry operation, for storage and distribution facilities; so potential savings must be carefully calculated and balanced against capital investment.

Many latex producers now offer higher solids latexes for use with pigment slurries.

TABLE 8.5. PRODUCTION TIME, DRY PIGMENT VS. SLURRY.

	With dry pigment, minutes	With pigment slurry, minutes
Charge liquid constituents	10	10
Add pigment	30	5
Disperse pigments	20	0
Pump out batch	10	10
Clean tank	20	20
	90	45

Slurry Properties

Titanium dioxide slurries are usually available at 64.5 ± 0.5% solids for flat systems and 76.5 ± 0.5% solids for gloss or semigloss systems. Slurries contain dispersants and biocides. The pH is 8.5–9.0, and viscosity is approximately 500 cp. (See Table 8.6.)

Slurries are substituted in formulas for equivalent titanium dioxide. However, the dispersant is reduced by about 20% in the formula.

COLOR PIGMENTS

Color pigments can be divided into two classifications, inorganic and organic, as shown in Tables 8.7 and 8.8.

Pigments are produced in a variety of physical forms to satisfy specific working properties or requirements:

Toners are unextended or full-strength organic pigments having the maximum tinting strength of their type.

C.P. (chemically pure) undiluted full-strength inorganic pigments are another type.

Resinated toners are organic pigments treated with barium or calcium resinate, which acts as a dispersion agent.

Lakes are pigments treated with alumina hydrate or some other transparent medium to give lower opacity but improve flocculation-resistant qualities and ease of dispersion.

TABLE 8.6. RUTILE TITANIUM DIOXIDE SLURRIES.

	For flat paints	For gloss and semigloss paints
Percent solids	64.5 ± 0.5%	76.5 ± 0.5%
Dispersant type	Inorganic-organic (nonphosphate)	Inorganic-organic (nonphosphate)
Biocide type	Nonmercurial	Nonmercurial
pH	9.0 ± 0.5	8.5 ± 0.5
Specific gravity	1.9 ± 0.5	2.35 ± 0.5
Weight per gal., lb.	16.1 ± 0.5	19.5 ± 0.5
Dry solids/gal.	10.4 ± 0.1	15.0 ± 0.1
Settling	Minor soft	Minor soft
Viscosity	Approx. 500 cp[a]	Approx. 500 cp[a]
Rheology	Newtonian (pumpable)	Newtonian (pumpable)
325 Mesh residue	0.02% max.	0.02% max.

[a]Brookfield viscosity (HBT 100 rpm #3 spindle).

Extended pigments comprise a physical mixture of prime pigments (organic or inorganic) with colorless extenders.

Easy-Dispersing Pigments. Another type that should be added to this listing is particle-size pigments (easy-dispersing). The trend over the last few years has been toward dispersing pigments in high-speed equipment to avoid long

TABLE 8.7. COLOR INORGANIC PIGMENTS

Blues	Oranges
Iron	Cadmium
Ultramarine	Chrome
	Molybdate
Greens	Reds
Chrome	Cadmium
Chrome oxide	Lead
Hydrated chromium oxide	Iron oxide
Yellows	Browns
Cadmium	Iron oxide
Chrome	Blacks
Iron oxide	Iron oxide
Nickel titanate	
Zinc	
Strontium	

TABLE 8.8. COLOR ORGANIC PIGMENTS.

Blues	Oranges
Copper phthalocyanine	Benzidine
Toners	Vat dye
Indanthrone	Dinitraniline
Greens	Reds and maroons
Copper phthalocyanine	Lithol
Toners	Para
	Toluidine
Yellows	BON
Nickel azo	Lithol rubine
Hansa	Quinacridone
Benzidine	Thioindigo
Vat	Arylide
	Violets
	Carbazole
	Toners
	Mincral
	Quinacridone
	Blacks
	Carbon

milling times. With particle size broken down by physical or chemical means to its maximum fineness, fewer problems will be encountered by over- or undergrinding. Great strides have been made by the pigment industry toward producing better, more durable, and more easily dispersing pigments. This trend will continue until practically all pigments will be of stir-in types.

Micronized Pigments. Pigments such as iron oxides are usually ground in disintegrators, roll mills, or pin mills. A small proportion of pigment aggregates and pigment agglomerates are left over. In the production of paints these agglomerates are not dispersed to sufficient fineness in sand mills or bead mills. By means of micronizing, a special grinding method using steam, the agglomerates are practically completely broken down, while the primary particles of the pigment remain preserved. The color shade of micronized pigment is no different from that of normally ground pigments. Only the tintorial strength increases by about 5%. Micronized pigments have a greatly reduced residue on the 325 mesh sieve (from 0.05% for normal to less than 0.001%) and a considerably improved ease of dispersion.

TINTING STRENGTH AND DISPERSION PROPERTIES

Tinting strength, the tintorial property of a pigment, is defined as the degree to which the colorant imparts color to a standard white pigment. It corresponds to the color yield of a pigment. Ease of dispersion is a measure of the time it takes for a colorant to reach maximum strength. The tinting strength of a pigment, in a set processing time, is dependent on its ease of dispersion; the harder the pigment, the weaker the apparent strength, and the softer the pigment, the greater the strength.

INORGANIC PIGMENTS

Chrome Yellows

Chrome yellows are forms of lead chromate. The palest yellows are obtained by coprecipitation with lead sulfate. These pigments are low in cost, have clean color and high hiding, and are nonbleeding and easy to disperse. They have poor alkali and soap resistance and tend to darken on exterior exposure in masstone. Since they contain lead, they darken on exposure to atmospheres containing sulfur. They can be used in waterborne paints, but may seed and produce low gloss.

Zinc Yellows

Zinc yellows, which are zinc chromates, are weak yellows. The pigment is a corrosion-inhibiting pigment because of the chromate ions released. Zinc chromate has poor color and hiding power, and poor acid and alkali resistance. Its principal use is in primers. It can be used in emulsions and will improve salt spray on bare steel.

Strontium Yellows

Strontium yellow is a precipitated strontium chromate. It is nonbleeding and easy-grinding and has good lightfastness in both masstones and tints, as well as good heat resistance. Its disadvantages are its very low tint strength and poor acid and alkali resistance. It can be used in waterborne paints.

Nickel Titanate Yellows

Nickel titanate yellow is commonly referred to as sun yellow or titanium yellow. This pigment is composed of oxides of nickel, antimony, and titanium. It is weak in strength, but has very good lightfastness, good outdoor weathering durability, and good chemical resistance, and can withstand high temperatures without color change.

Cadmium Yellows

Cadmium yellows are nonbleeding and have excellent lightfastness in deep shade, excellent baking resistance, and good alkali resistance. Their disadvantages are low tinting strength, poor gloss retention, and high cost.

Yellow Iron Oxides

Yellow iron oxide is a relatively low-cost pigment that has good lightfastness in both masstone and tints, is nonbleeding, is free of toxic materials, and has high hiding power and good chemical resistance. Its disadvantages are that it is dirty in color, has a tendency to settle, and has poor gloss retention. The pigment is available as a synthetic or in its natural state. It is used in water-soluble-type paints.

Chrome Oranges

Chrome oranges are low in cost and nonbleeding, and have good lightfastness. They contain lead, darken in masstone, have poor alkali and soap resistance, and have poor hiding in comparison to molybdate oranges.

Molybdate Oranges

Molybdate oranges are low in cost, nonbleeding, lightfast, and easy to disperse; they also have good bake resistance. Their disadvantages include the presence of toxic metals, poor alkali and soap fastness, and a tendency to darken in masstone on exposure. The pigment is used extensively in blending with organic red pigments to produce low-cost, high-intensity red finishes that have good hiding and lightfastness. This pigment can be used in water solution and dispersible-type paints.

Cadmium Oranges

Cadmium orange, like cadmium yellow, is available in both C.P. and lithophone. It is basically the same as cadmium yellow—high in cost and low in tint strength. The orange hue is obtained by introducing cadmium selenide into the pigment.

Chrome Greens

Chrome greens are coprecipitates of chrome yellow and iron blue. They are low in cost, have high hiding and good bake resistance, and are nonbleeding. Their disadvantages are poor alkali resistance, moderate lightfastness, and a tendency to flood and float. This pigment fades and is reactive in waterborne coatings.

Chromium Oxides

Chrome oxide is a calcined inorganic green pigment that has good heat, light, and chemical stability. It is easy-grinding, but has poor tint strength. It is recommended for water-soluble-type paints.

Hydrated Chromium Oxides

Hydrated chromium oxide, also known as C.P. permanent green, is quite different from chromium oxide. Hydrated chromium oxide is prepared by calcining bichromate of soda in the presence of boric acid, with subsequent hydrolysis during washing. It is clean in color and high in chroma, and has excellent outdoor durability. It is also used for tinting pastels because of its excellent color stability. Its disadvantages are low tinting strength, difficulty of dispersion, and poor blister resistance.

Iron Blues

Iron blues have high tinting strength, are low in cost, and have good durability and lightfastness in masstones and dark tints. Common names are Chinese, Milori, and Prussian Blue. Their disadvantages are poor lightfastness in light tints, poor storage stability, and poor alkali resistance. They are not recommended for waterborne coatings.

Ultramarine Blues

Ultramarine blue is a deep reddish blue made by firing a combination of sulfur, alkali, clay, and a reducing agent at elevated temperatures. It is low in cost and has good alkali resistance and excellent heat resistance. Its disadvantages are very poor acid resistance, low hiding power, poor tinting strength, poor outdoor durability, and poor wettability. It performs poorly in high-pH systems.

Red Iron Oxides

Red iron oxides have relatively the same characteristics as the yellow iron oxide, and can also be obtained in synthetic or natural form. Pure red oxides in synthetic form are easy to disperse. The synthetic form is carefully standardized for color, tinting strength, hiding power, ease of dispersion, particle size, oil absorption, composition, and many other properties that are essential to quality paint performance.

Natural types of iron oxides are lower in price and are used in floor enamels, primers, and barn paints. Synthetic iron oxides are suitable for water solution and dispersible paints. Avoid natural iron oxides that contain calcium sulfate impurities.

Red Leads

Red lead is a very old pigment, low in hiding, nonbleedable, and easily dispersible. It is very high in specific gravity and low in oil absorption. This pigment is used primarily for undercoats on iron structures because of its rust-inhibitive properties. It can be used in waterborne coatings, but may seed and settle.

Cadmium Reds and Maroons

Cadmium reds have the same characteristics as the cadmium yellows except for durability, which is much better. The reds are coprecipitated. Cocalcined mixtures of cadmium sulfide and cadmium selenide, those that contain quantities of barium sulfate, are known as cadmium lithophone. These pigments are high-hiding, nonbleeding, very lightfast in deep shades, and very bake-resistant. Their disadvantages are low tint strength, poor gloss retention on outdoor exposure, and poor lightfastness in tints. They can be used in water-soluble-type paint but may develop low gloss.

"Mercadium" Reds and Maroons

"Mercadium" reds and maroons are calcined pigments made by combining sulfides of mercury and cadmium. They have properties about equal to those of the cadmium, being slightly brighter in masstone and cleaner in tints.

Siennas, Ochres, and Umbers

Siennas, ochres, and umbers are all earth colors. Iron oxide in various forms makes up these colors, which are permanent to light in both masstones and tints. They do not bleed in water, oil, or solvents. Colors are resistant to alkali, but not to acids.

ORGANIC COLOR PIGMENTS

Nickel Azo Yellows

Nickel azo yellows, commonly referred to as green gold, have lightfastness in masstone and tints, very slight bleed in some vehicle systems, and good bake resistance.

Hansa Yellows

Hansa yellow, commonly known as toluidine yellow, is an insoluble azo organic pigment. It is excellent in masstone lightfastness, has good alkali and acid resistance, is lead-free, and is easy to grind. Its disadvantages are poor lightfastness in tints, bad bleeding in almost any type of vehicle, relative transparency, poor heat resistance, and high price. It can be used in waterborne paints.

Benzidine Yellows

Benzidine yellows are insoluble azo organic pigments having approximately twice the tinting strength of Hansa yellow. The *o*-toluidine type is used in the paint industry and has high tint strength, is free of toxic metals, and has good masstone lightfastness and soft texture. These pigments also have good chemical resistance, excellent bake resistance, and good bleed resistance. Their disadvantages are high cost, low hiding power, and high oil absorption. They are suitable for emulsion paints.

Vat Yellows

Vat yellows are commonly referred to as anthropyrimidine and flavothrone yellows. They are alkali- and acid-resistant, are nonbleeding, and have excellent exterior durability. They are high in cost in both toner and lake form.

Benzidine Oranges

Benzidine orange, commonly referred to as pyrazolane orange, is a diazo-pigment dyestuff prepared by coupling diazotized dichlorobenzidine with two molecules of phenylmethyl-pyrazolone. It has good chemical resistance and is easy to disperse. Disadvantages include high cost, poor lightfastness, poor bleed resistance, and low hiding power. It is satisfactory for emulsion paints.

Dinitraniline Oranges

Dinitraniline orange is an (mono) azo pigment dyestuff prepared by coupling diazotized dinitroanaline to β-napthol. It has high tint strength, is free of toxic metals, and has good chemical resistance and good masstone lightfastness. Its disadvantages are high cost, poor lightfastness, and poor bleed resistance. It is used in waterborne paints.

Vat Dye Oranges

Vat dye oranges are commonly referred to as orange RK or "indofast" orange. They are anthraquinone derivatives. They have excellent lightfastness in pastels, freedom from toxic metals, excellent chemical resistance, and excellent bake resistance. They are very high in cost in both lake and toner types and have a tendency to show slight bleed in some systems.

Copper Phthalocyanine Greens

Phthalocyanine green is derived from phthalocyanine blue by chlorination. Chlorination, in addition to producing a green pigment, imparts still greater stability to the molecule. It is available in a number of forms: full-strength toner, resinated toner, flushed form, water-dispersible form, and reduced form. It is clean in color (high chroma) and has high transparency for metallics, excellent chemical resistance, bake resistance, lightfastness, tinting strength, and bleed resistance. However, it is high in cost, except in tints, and dark shades have a tendency to bronze on exposure. This pigment is seldom, if ever, used as a straight masstone pigment. It is used in waterborne paints.

Organic Green Toners

Organic green toners are commonly referred to as PTA (phosphotungstic) or

PMA (phosphomolybdic) green. The PTA types are considered to have better brilliancy and slightly better lightfastness than PMA. Both types are transparent in masstone, very high in tint strength, and high in chroma. Their disadvantages are high cost, poor outdoor durability, and poor bleed resistance.

Copper Phthalocyanine Blues

Phthalocyanine blues have excellent lightfastness, heat stability, and chemical resistance. They are very high in tint strength and have excellent bleed resistance, but are high in cost. They are recommended for waterborne paints.

Organic Blue Toners

Organic blue toners are similar in properties to organic green toners.

Indanthrone Blues

Indanthrone, or "indo" blue is classified as a vat dye pigment. It has outdoor durability in all depths of shades, good chemical resistance, good bake resistance, and good bleed resistance. Indanthrone blue is considerably redder in hue than phthalocyanine blue and is far superior in bronze resistance.

Carbazole Dioxazine Violets

Carbazole dioxane violet is known as "indofast" violet, hostaperm vat violet, monarch violet, carbazole violet, and so forth. It is an opaque organic toner with excellent lightfastness, bleed resistance, and alkali and acid resistance. However, it is high in cost and difficult to disperse.

Organic Violet Toners

Both PTA (phosphotungstic) and PMA (phosphomolybdic) pigments are used in the printing ink industry. They have high tinting strength and high color purity. Their disadvantages are high cost, poor outdoor durability, and poor bleed resistance.

Mineral Violets

Mineral violet, an organic pigment, is referred to as manganese violet. It has excellent bake resistance, excellent acid resistance, excellent bleed resistance, and is low in cost. Its disadvantages are very low tinting strength, poor alkali resistance, hard grinding, and poor suspension. The color has excellent lightfastness for interior exposure, but is not recommended for exterior use.

Quinacridone Violets

Quinacridone violet, commonly referred to as "quindo" and "monastral violet," is alkali- and acid-resistant, nonbleeding, lightfast in masstone and tints, and heat-resistant; it also has high hiding power. These pigments are high in cost and are used extensively in combination with molybdate orange to give brilliant low-cost reds with durability.

Lithol Reds

Lithol reds are the most economical organic reds available, have good bleed resistance, and are easy-grinding. They are widely used where brilliance, hiding, and low cost are of primary importance, but they are recommended for interior use only. The acid resistance of barium and calcium lithols is generally considered good; alkali resistance is only fair. The sodium lithols have poor alkali and acid resistance.

Para Reds

Para reds are relatively low in cost, with good hiding power, ease of grind, good acid and alkali resistance, and good deeptone lightfastness. They are not recommended for tints because of very bad bleed and poor lightfastness, or for baking finishes where they brown with heat.

Toluidine Reds

Toluidine reds are exceptionally bright reds produced in two shades—light and dark. They are relatively low-priced, are easy-grinding, and have excellent deeptone, lightfastness, and outdoor durability; they also have excellent acid and alkali resistance, but very poor bleed resistance. They can be used in waterborne paints, but may haze where glycol ethers are present. They can also contribute to loss of dry or viscosity instability.

BON (B-Oxynaphthoic) Reds and Maroons

BON reds and maroons are commonly referred to as rubine reds and are chemically similar to lithol rubine, being treated with calcium or manganese to improve masstone lightfastness. These pigments range in shade from relatively light red to very dark maroon. They have very good masstone, lightfastness and excellent bleed resistance. Their disadvantages are poor alkali and soap resistance and poor tint lightfastness. They are used in water-dispersion systems, but may fade at high pH.

Lithol Rubines

Lithol rubine is a deep masstone red (high chroma) with poor hiding power. It is used in blending with molybdate orange for low-cost reds. Disadvantages are poor lightfastness and poor alkali and soap resistance. It can be used in waterborne systems, but may fade at high pH.

Chlorinated Para Reds

Chlorinated para red, commonly known as tanager, fire toner, blazine red, permaton red, and so on, is closely related to para red. Its properties are similar to para red except that it has better tint lightfastness. It is used in latex paints.

Quinacridone Reds and Maroons

Quinacridone reds and maroons are commonly known as MONASTRAL[(R)] and quindo colors. They have exceptional color brightness, outstanding lightfastness in tints, and excellent resistance to chemicals, acids, alkali and soap. They are nonbleeders and have excellent resistance to heat, but they are expensive. They can be used in waterborne paints.

Thioindigo Reds and Maroons

Thioindigos are commonly referred to as "indo" or "thiofast" pigments. They have excellent chemical resistance, excellent lightfastness in dark shades, and good bake resistance. They are high in cost, bronze in masstone, and bleed in some instances.

Arylide Maroons

Arylide maroons, commonly referred to as toluidine maroon, "Naphthanil[(R)]," and naphthol maroons are azo pigments based on arylides of B-hydroxy-naphthoic acid. These colors have a yellowish undertone. These pigments bleed severely and are very hard to disperse. Lightfastness in masstone depth is good, and they have excellent acid and alkali resistance. They are low in hiding power and are relatively expensive.

[(R)]Trademark—E. I. du Pont de Nemours & Co.

Carbon Blacks, Lampblacks, and Bone Blacks

Carbon blacks are manufactured from raw gas, whereas lampblacks are made from oil. Bone blacks are made from cattlebones which are charred and then ground to pigment fineness. Jetness or color intensity of blacks are measured on an instrument called a Nigrometer. This instrument gives a numerical number to each black—the higher the number is, the blacker the color. For example, a carbon black in the 255 range would be considered a very jet black, whereas a black in the 60 range would be considered a gray. The smaller the particle size, the blacker the color.

PIGMENT DISPERSION

The key to good pigmented coatings is proper dispersion. The natural tendency of pigment particles is to flocculate because small particles have large interparticle attractive forces working between them. In aqueous media this flocculating tendency is intensified when pigment particles are squeezed together by water molecules. Water has a strong affinity for itself, a consequence of its high cohesive energy density.

In formulating, these natural tendencies must be permanently overcome if pigments are to remain dispersed in the liquid coating. The dry coating should have discrete pigment particles evenly distributed throughout to have maximum durability, hiding, color, and gloss properties.

In the dispersing process, the air from the particles is displaced with the liquid grinding medium containing the dispersing agents. These surface-active agents are absorbed onto the surface of the pigments, thus preventing flocculation and aggregation by lowering the surface energy of the liquid and improving the wetting process.

Defoamers, especially those based on silicones, should be incorporated at this stage to be most beneficial by being absorbed onto the pigment surface. If the silicone defoamers are not absorbed by the pigments, they may cause cratering in the finished product if not properly dispersed.

The effectiveness of a dispersing agent is dependent upon its inherent ability to be absorbed onto pigment particles. Another factor affecting dispersing efficiency is whether the pigments are organic or inorganic. The type of surface treatment on the inorganic pigment also has a great influence.

There is a material difference between dispersants and dispersing resins. Dispersants have relatively low molecular weights, but large numbers of functional groups per molecule, with one of the most effective being the carboxyl group. Therefore, dispersants act mainly through electrical phenomena and are most efficient with waterborne coatings.

Dispersing resins have higher molecular weights and fewer functional groups, are generally less efficient, and must be used at higher concentrations.

The advantage of dispersing resins is that they become an integral part of the binder network, thus making the coating less water-sensitive.

Dispersing agents have a balanced affinity for both the pigment surface and the liquid phase. They also can stabilize the system by making it mutually compatible. They function through one or a combination of the following factors: ionic charge, hydrogen bonding, dipole–dipole interaction, and protective colloids.

Ionic charge is the most powerful mechanism for stabilizing a pigment dispersion. In an aqueous system, the anionic portions of the ionic dispersant absorb preferentially onto the pigment surface, thus forming a negatively charged layer at the pigment–liquid interface. The anions attract positively charged ions in solution. The resulting electric double layer is effective in causing repulsion of the pigment particle.

The electrical balance in some ionically dispersed systems can be upset by either excess dispersant or the addition of inorganic salts. Flocculation, described as "salting out," usually occurs. The inorganic phosphates are an example, and are particularly susceptible to salting out.

Hydrogen bonding helps stabilize a pigment by forming a buffer around the pigment particle. Hydrogen bonds result by orientation of the water molecule with the adjacent molecules because of its electropositive and electronegative regions. Hydrogen bonding is also responsible for viscosity increases.

Dipole dispersing agents do not ionize in solution. They have definite negative and positive poles. When one pole is on the pigment surface, the opposite pole extends into the liquid phase, thereby repelling pigment particles similarly protected.

Protective colloids, which are used in emulsion systems, function through their size. They are usually of high molecular weight and form a physical barrier by absorbing onto the pigment particles and hindering particle-to-particle contact.

The proper choice of a dispersant or blend of dispersants is based upon its function and expected performance. The most important criterion in the selection of a dispersant should be the HLB value (see Chapter 6). Each pigment or combination of pigments requires a certain amount of dispersant or combination of dispersants. A dispersant may have the correct HLB, but be of the wrong chemical family for use with a specific pigment. For a given dispersing system, there is a range of effective concentration that will stabilize the system.

Concentration-Aggregation Method for Determining Dispersant Demand

One method for determining the dispersant demand of pigments dispersed in water is the concentration-aggregation method.* In this procedure, 50 g of

*Rohm & Haas method.

pigment or extender are mixed with sufficient water to form a stiff, slightly moist mass. Pigments may be run individually or blended in the rate used in a paint formulation. The dispersing agent, 10% water solution, is then added in small increments; agitation follows each addition when the pigment mass becomes fluid enough so that surface ripples disappear on gentle shaking of the container. The pigment system is tested for deflocculation as follows:

A small portion (about 1 cc) is placed on a watch glass. To this, a 6% solution of an anionic thickener is added dropwise. Flocculation will occur if insufficient dispersant has been used. If flocculation occurs, more dispersant is added to the main pigment mass and another fraction tested. These steps are repeated until samples show no tendency to flocculate. Empirical work indicates that excellent paints can be made using about one-third of the amount of dispersant determined by the concentration-aggregation method.

Table 8.9 lists the dispersant demands of some pigments and extenders.

TABLE 8.9. DISPERSANT DEMANDS OF PIGMENTS AND EXTENDERS. (% DISPERSANT BASED ON DRY PIGMENTS*).

Barytes (fine particle size)	0.32	Rutile titanium dioxide	2.4
Barytes (medium particle size)	0.48	Anatase titanium dioxide	3.5
Calcium carbonate (2.5 microns)	0.24	Yellow iron oxide	1.4
Calcium carbonate (5.0 microns)	0.40	Brown iron oxide	0.82
Calcium carbonate (14 microns)	0.20	Black iron oxide	4.9
Calcium carbonate (precipitated)	0.30	Red iron oxide	2.9
Calcium silicate	0.82	Chrome oxide	1.7
Clay (1 micron)	2.2	Sun yellow	0.4
Clay (5 microns)	2.0	Lamp black	1.3
Mica	1.8		
Silica	0.2		
Talc	1.0		

*Dispersant solution 1:1 mixture of Tamol 731 and Triton CF-10 at 10% solids.

9

Biocides—Preservatives and Fungicides

The term "biocides" includes both bactericides and fungicides used in waterborne paints to prevent microbial growth in the packaged paint and the growth of fungi on the dried paint film.

Bacteria and fungi require moisture, nutrients, and the proper temperature to exist. Under these conditions and in the absence of inhibiting substances, they will multiply, with the final end result being the destruction of the paint.

PACKAGE PRESERVATIVES

In an aqueous paint, the protective colloids (casein, soya protein, alginates, starches, natural gums, and cellulosic derivatives), the latex or binder, and the fatty emulsifiers or defoamers can all be nutrients for the microorganisms.

When an aqueous paint is contaminated with microorganisms and growth conditions are favorable, enzymatic action begins. This may result in viscosity breakdown, gassing, obnoxious odors, bulging of the container, a pH drift, and a general breakdown of the system. In solvent paints, bacteria attack is not usually a problem, since these microorganisms cannot exist in the absence of water. Also, the solvent vapors in the air space of the can act as vapor retarders. Fungi attack a dried film forming a dark deposit (sometimes mistaken for dirt collection), and this leads to premature breakdown and loss of adhesion of the coating. (See Figures 9.1 and 9.2.)

Bacteria are minute living cells barely visible under a microscope. Microorganisms require moisture, usually above a 35% moisture content in the film, and a nutrient which normally contains carbon. However, some bacteria and fungi are capable of utilizing sulfur, iron, and the like, directly as an energy source. Some other factors that affect microbiological development are temperature, availability of oxygen, pH, surface tension, and presence of inhibiting substances. The ideal temperature for bacterial development is 77–99°F and for fungi is 72–86°F.

Most growing organisms require large amounts of oxygen to survive; this type of organism is called aerophilic. Another type of bacteria which does not require oxygen is known as anaerobic. These bacteria can live deep in liquid

FIGURE 9.1. Mildew on painted wood surfaces.

FIGURE 9.2. Mildew on wood sliding—portion washed to remove mildew.

systems without dissolved oxygen or in closed containers. Most microorganisms prefer a neutral system (pH 7) but some can tolerate pH ranges from 3 to 10.

Microorganisms produce enzymes which are true organic catalysts; they are capable of functioning independently of the cells that produce them. They are large protein molecules that will degrade a large variety of compounds. For instance, cellulose, which is water-insoluble, is attacked by enzymes and degraded to cellubiose and then to glucose. Glucose, which is water-soluble, can then be utilized by the bacteria.

The first part of the problem we are concerned with is the preservation of the paint in the package. Since microorganisms can be introduced into the system through raw materials, it is important to protect the system with a preservative as soon as possible. For instance, bacteria may be introduced with the system in an extender pigment such as a calcium silicate which is mined from natural sources. When introduced with the paint mix, the microorganisms start to multiply and attack organic matter in the paint system.

Snyder and Williams[4] report that they have isolated a great many bacteria from contaminated dispersed pigments, raw materials, and paints. Tentatively, they have been identified as *Pseudomonas sp., Aerobacter aerogenes, Escherichia coli, Bacillus subtilis, B. mycoides, B. cereus, Micrococcus flovus, Achromobacter sp.*

Usually a bacteria count will tell if the paint has been "spoiled" by bacteria attack. However, in cases where a batch of paint may stand two or three hours before the addition of the preservative, the bacteria, if present, will have produced enough enzymes to be capable of degrading the thickener. When such a batch, low in viscosity, is checked for bacteria count, the count may be zero if enough preservative had been added. Based on these results, the low viscosity would not be attributed to bacteria. This conclusion would fail to take into account the enzymes produced before the batch was sterilized. If such a batch were then mixed with a high-viscosity batch, the combined material would lose viscosity in a short time.

Proteins employed as emulsion stabilizers are hydrolyzed to amino acids, and further degradation of these amino acids produces putrefactive odors. Carbohydrates used as stabilizers are reduced to simple sugars by bacteria enzymes as indicated by chromatographic studies.

Growth experiments indicate that styrene-butadiene, polyvinyl acetate, and acrylic resins can serve as a sole source of carbon in an otherwise complete growth medium for *proteus sp., Aerobacter sp.,* and *Flavobacterium.* This work of Ross and Buckman[3] shows the sharp decreases in viscosity of the latices due to attack from microorganisms.

Resin	Microorganism	Viscosity cp
Styrene-butadiene	Uninoculated	750
Styrene-butadiene	*Proteus sp.*	325
Styrene-butadiene	*Aerobacter sp.*	310
Styrene-butadiene	*Flavobacterium*	355
Polyvinyl acetate	Uninoculated	800
Polyvinyl acetate	*Proteus sp.*	250
Polyvinyl acetate	*Aerobacter sp.*	220
Polyvinyl acetate	*Flavobacterium sp.*	285
Acrylic	Uninoculated	9.5
Acrylic	*Proteins sp.*	4.6
Acrylic	*Aerobacter sp.*	5.4
Acrylic	*Flavobacterium*	6.0

Spoilage may sometimes produce strings of microbiological growth, commonly called slime, which do not materially affect the properties of the paint in the can but result in streaks when the paint is applied.

A desirable preservative product should have a high antimicrobial efficiency and a broad spectrum of antimicrobial activity. A preservative should have no undesirable effects such as discoloration on the film characteristics. The preservatives most commonly used are mercury compounds and phenol compounds, although others may be employed.

For preservation in the package, mercury, in the form of phenyl mercuric acetate, is used in the amounts of 0.03 to 0.08% of the total wet weight of the paint. Phenol-type products are used in the concentration of 0.6 to 0.8% of the total wet weight of the paint. "Dowicide*" A and "Dowicide" G are commonly used for this purpose.

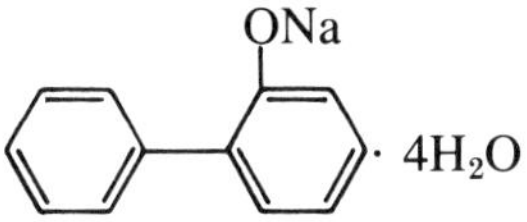

ONa · $4H_2O$

Dowicide A
O-Phenylphenol Soidum salt

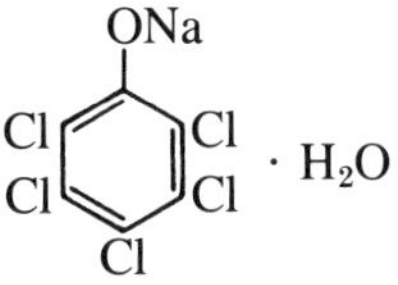

ONa
Cl Cl Cl Cl Cl · H_2O

Dowicide G
Pentachlorophenol Sodium salt

Preservatives must be added carefully to the paint. Some are in the dry form and should be dissolved in water before their addition to the paint. As mentioned before, they should be added to the mix as soon as possible to give protection during the manufacturing process. If added too quickly without proper agitation, they can cause coagulation of the latex or flocculation of the pigment.

Reviewing again the procedures for keeping microorganisms under control in paint manufacture:

1. Maintain a constant vigil on raw materials, storing them properly.
2. Use clean equipment in manufacture, which means in some cases cleaning down to bare metal.
3. Introduce preservatives into the mix as soon as possible.
4. Do not blend off spoiled batches with good batches.
5. Keep airborne dust to a minimum.

ENZYME RESISTANCE

Latex paints containing cellulosic thickeners will lose viscosity during the storage if the paint has been contaminated with cellulolytic enzymes. These enzymes are extracellular proteins produced by certain species of bacteria,

*Trademark for Dow Chemical Co.

fungi, and yeasts that catalytically cause the polymer chains of the thickener to be cleaned at adjacent positions of unsubstitution. The resulting viscosity loss is due largely to the thickener's inability to continue to efficiently bind the water in the vehicle because of its greatly reduced chain length.

Although the "in-can" preservatives in the formulation may sterilize the paint for 4 to 48 hours after incorporation, there are no known preservatives that can denature or permanently render cellulolytic enzymes inactive; also, the concentration of enzymes produced by microorganisms during storage prior to sterilization may not be significant. Most of the problem of enzyme contamination occurs from the use of contaminated raw materials such as latex, thickener solutions, aqueous colorants, and pigment slurries. Test procedures are available for detecting enzymes in raw materials.* Enzymes can be introduced into paint through poor plant housekeeping. Enzyme activity is most prevalent at pH between 3.0 and 9.0. Hydroxyethyl cellulose resistance to enzymes has been improved by changing the distribution of the hydroxyethyl groups in the molecule.

A method† has been developed for measuring the enzyme resistance of paints. Samples of latex paints are inoculated with a fixed amount of an aqueous, fungal enzyme solution of known concentration. The paint viscosity is then measured periodically over a period of six months. The test may be terminated whenever one of the paints loses 20% of its original viscosity. Percent viscosity retained is reported:

$$\text{percent } N_R = \frac{N_t}{N_o} \times 100$$

where percent N_R = percent viscosity retained after time t.
N_t = final paint viscosity at time t.
N_o = initial paint viscosity.

FUNGICIDES

The second type of attack of microorganisms is on the dried paint film. Here molds and fungi damage and discolor the coating.

Moisture, temperature, and nutrition are again the three general requirements for mildew growth on paint. Moisture contents of 10–20% or relative humidities of 70–75% are sufficient. The ideal temperature for fungoid growth is 72–86°F, although they will grow outside of this temperature range.

Nutrients are present in the paint film and can also be deposited on the surface from airborne dust.

Mildew growth is no more common on paint than it is on many other

*Hercules Bulletin VC 492, *Test Procedure for Detecting Enzymes in Latex Paint.*
†Hercules Incorporated.

materials. Other conditions being equal, it grows on soft paints more readily than hard ones. Heavy chalking paints are more resistant than chalk-resistant counterparts; they are truly self-cleansing.

Mildew growth may be mistakenly identified as dirt collection. Examination under a microscope will show the characteristic fungi pattern. (See Figure 9.3.) If the discoloration is fungi, a drop of commercial sodium hypochlorite solution (5% active chlorine) will bleach out the color.

Undercoatings play an important role in the growth of molds on paint films. If a latex paint is coated over wood, for instance, it is more susceptible to mold growth than if coated over a cement board.

Before painting a mildewed surface, it is necessary to remove the mildew. Dry wiping and scraping are usually ineffective. Wash with mild alkali, such as washing soda or trisodium phosphate. If the mildew has penetrated the surface, an abrasive cleaner is better, If any mildew remains, the surface should be washed with a bleach solution.

Ross[3] and Buchman report the following:

Microorganisms isolated from emulsion paint films exposed on test fences at six different locations in the United States.

Fungi	*Bacteria*
Alternaria dianthicola	*Bacillus mycoides**
Aspergillus flavus	*Flavobacterium marinum**
Cephalosporium carpogenum	*Sarcina flava*
Cladosporium sphaerospermum	
Cladosporium sp.	
Helminthosporium speciferium	
Penicillium oxalicum	
*Phoma glomerata (Peyronillaea)**	
*Pullularia pullulans**	
Stemphylium consortiale	
Torula nigra	
Unknown yeasts	

*Isolated in sufficient numbers to be considered the normal microflora of paint films.

Not all of the molds were encountered in each geographical area. *Pullularia pullulans* was observed more frequently than the others in all areas except Delaware and New York. In these areas another black mold, later identified as species of *Cladosporium,* was observed as frequently as *Pullularia pullulans*. However, the two molds are similar in that they are both typically black friable structures and approximately the same size. Because of these similarities, it is difficult to differentiate the two, and a casual examination might lead to the erroneous conclusion that both are *Pullularia*.

Approximately one-third of the observable disfigurement of panels examined in California was the result of growth of *Phoma glomerata*.

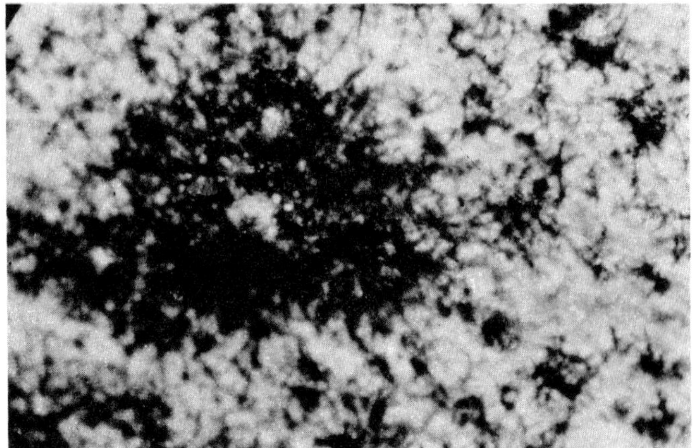

FIGURE 9.3. Microphotograph of fungi. 40×

The greatest variety of microorganisms was isolated from panels examined in Miami. The favorable growth conditions of that area make this finding unsurprising.

Mildew may be black, green, red, purple, or gray, but the species of *Pullularia, Alternaria,* and *Phoma* are usually black.

Dresher[1] describes the isolation and identification of microflora on exterior emulsion paints. Krumperman[2] describes the microorganism found on the interior paint films of food processing plants.

It is believed that the principal way that biocides kill microorganisms is probably by coagulating the protein portion of their protoplasm. Biocides may also increase the hydrogen ion concentration around the cell, which creates unfavorable conditions, since a neutral (pH 7.0) environment is preferred. They may also lower surface tension, which disrupts cell membranes and cuts off food supply by blocking enzymes, thereby interrupting metabolism. While emulsion paints require preservatives to keep microorganisms under control in the package, water-soluble-type paints require little or no peservative. It is believed that the presence of the cosolvent inhibits the microorganisms: however, the dried paint film in the paint-production area may be susceptible.

Woods such as cedar are quite mold-resistant, while paints applied over pine show varying degrees of mildew resistance.

Areas on the exterior of homes where the painted surfaces are shaded by trees or shrubbery so as to retain moisture longer are more susceptible to mildew attack than less moist areas.

Mildew infestations vary from year to year, depending on climatic conditions. Mildew may suddenly appear in areas where it has never been known to be present before.

Government recommendations have restricted mercury compounds in paints to use as a preservative for interior waterborne paints and as a fungicide for exterior waterborne paints. The Federal Hazardous Substance Act limits the use to 0.2% mercury (calculated as metal) based on the total weight of paint. The current OSHA standard (TLV) for inorganic and phenyl-mercury compounds indicates a level of 150 micrograms per cubic meter.

With restrictions on mercury compounds, many nonmercurial mildewcides are used. Mercury fungicides are still the lowest in cost and effective over a long period of time in latex paints. The cost of fungicide at the present time is about 12¢ per gallon, but could go as high as 60¢ per gallon in hot, humid regions.

A list of nonmercurial fungicides is shown in Table 9.1. The Zinc Institute evaluated a series of fungicides used in combination with a series of fungicides. The most effective fungicides were:

2-*n*-octyl-4-isothiazolin-3-one
zinc dimethylithiocarbamate
diiodomethyl *p*-tolysulfone
2,4,5,6-tetrachloroisophthalonitrile
2,3,5-trichloro-4 (propylsulfonyl) pyridine
bromo-5-chlorobenzoxazolone

Zinc oxide has been used for many years in both oil-based and waterborne coatings as a mildewstat. It is not an active mildewcide; but when used as a part of the coating pigmentation, it usually has a definite inhibiting action on mildew growth. However, its effectiveness can vary substantially, depending on the coating composition and the exposure location.

The mechanism of the mildewstatic action of the zinc oxide is not clearly understood. There is considerable evidence to indicate that zinc oxide may be effective in controlling mildew by insolubilizing the enzymes produced by mildew to break down organic matter and provide food for the mildew growth. Another possibility is that zinc oxide may prevent germination of mildew spores.

Mildewcides are the major weapon in the fight against mildew on paint films. When properly used, they provide great improvement in the mildew resistance of protective coatings, particularly in the early part of a coating's life before the onset of chalking. The amount of mildewcide added in the coating formulation and its overall rate of depletion from the applied coating largely determine the length of time the mildewcide will provide protection against mildew. There are also considerable differences in efficiency of mildewcides against specific organisms.

In certain cases an unusual type of synergistic action is observed between mildewcide and zinc oxide. For example, Skane M-8 functions most effectively in combination with 25 to 50 pounds of zinc oxide per 100 gallons of coatings.

TABLE 9.1. NONMERCURIAL FUNGICIDES.

Trade name	Compound	Remarks
Cosan P Fungitrol 11 Advacide TMP	*n*-Trichloromethyl thiophthalimide	Questionable stability to hydrolysis.
Vanacide PA	*Trans* 1,2, bis (*n*-propylsulfonyl)ethene	Poor hydrolysis resistance.
Biocide DS-2787	2,4,5,6 Tetrachloro-isophthalonitrile	
	Zinc oxide	Use 0.5 to 1.0 lbs. per gallon. Some latices not stable in presence of zinc oxide.
Metasol Q	Copper-8-quinolinolate	Colored, increases chalk.
Busan M-1	Barium metaborate	Useful in some applications.
Skane M-8	2-*n*-octyl-4-isothiazolin-3-one	
Metasol TK-100	2-(4-thiazolyl)benzamidazole	

The ingredients in the coatings control the susceptibility of paint film to mildew growth. Oleoresinous modifiers are very susceptible to mildew attack. Calcium carbonate is a preferred extender for good mildew resistance because it maintains the coating at an alkaline pH and because it has a low binder demand and helps produce a tight coating. Materials that are water-sensitive, such as surfactants and thickeners, should be kept to a minimum to maintain maximum water resistance.

Complete data on the toxicity of fungicides must be supplied to the Environmental Protection Agency before it will grant registration of the product for use as a paint mildewcide. Some of the toxicity tests are:

Acute oral toxicity (rat LD50)
Acute dermal toxicity (rabbit LD50)
Primary dermal irritation (rabbit)
Primary eye irritation (rabbit)
Acute inhalation (rat LD50)
Skin sensitization (guinea pig)
Ames mutagenicity

There are many laboratory tests for predicting field performance of mildewcides in paints using Petri dish, filter paper, and environmental chamber tests.

REFERENCES

1. Drescher, R. F., *Am. Paint J.*, **42,** No. 27, 80–102 (1958).
2. Krumperman, P. H., *Am. Paint J.*, **42,** No. 38, 72 (1958).
3. Ross, R. T., and Buchman, *Offic. Dig. Federation Paint Varnish Prod. Clubs,* **31,** 276 (1959).
4. Snyder, H. D., *Am. Paint J.*, **46,** 74 (1962).
5. *1976 Coating Biocide Guide,* "Modern Paint & Coatings," pp. 22–23, April 1976.

10

Miscellaneous Ingredients

Emulsion and other water-type paints contain many additives to improve their performance. Besides the pigment, vehicle, surfactant, and protective colloid, there may be coalescing agents, freeze-thaw stabilizers, buffers, driers, anti-rust agents, and so on.

Many of the additives put into an emulsion-type paint are present to improve the stability. Since these are colloidal dispersions of many types of suspended particles, they are sensitive to destabilizing influences such as temperature extremes, pH, soluble salts, and mechanical stresses. During manufacture, the paint must be protected against extreme shear stress, temperature extremes, and colloidal shock from the too-rapid addition of the various materials. The paint must be stable in the can, and cannot be contaminated by metal salts present from can corrosion. The paint must be stable in the hands of the consumer, who may subject the material to considerable abuse because of lack of understanding.

Some additives are put into water-type paints to modify the application properties such as flow, wet edge, and other properties.

Additives must be carefully evaluated, as excessive amounts will have detrimental effects on other properties of the paint. For example, large amounts of coalescing agents will shorten the package stability time.

FREEZE-THAW STABILIZERS

The water in waterborne paints will freeze at low temperatures. In water-soluble-type materials, freezing is not as destructive as in dispersion systems, although it will cause some pigmet agglomeration. In emulsion-type paints freezing can be much more destructive. In an oil-in-water-type emulsion or a latex paint, the external phase is water. Upon freezing, ice crystals are formed, which expand and subject the dispersed particles to high pressures, forcing the particles together, which may cause coagulation. On thawing, each of the various components reaches maximum solubility at various times, and coagulation may take place if the fusion temperature is reached before the protective colloid again becomes effective. Slow freezing is much more destructive than fast freezing, as larger ice crystals are formed.

A paint is tested for freeze-thaw stability by seeing how many freeze-thaw cycles it will go through without affecting the package condition. Most commercial paints will pass three to five freeze-thaw cycles.

A cycle consists of 16 hours at a temperature of 10°F followed by 8 hours at 75°F. The paint is examined for pigment flocculation, resin flocculation, and increase in viscosity.

While particle size, resin hardness, surfactant, and protective colloid have some effect on freeze-thaw stability, the simplest way of increasing resistance to freezing is to depress the freezing point of the water. This can be accomplished by the use of such water-soluble additives as ethylene glycol, propylene glycol, glycol ethers, and so on. The use of antifreeze may be complicated by the presence of water-soluble coalescing agents which change the glycols to latent solvents. Many nonionic emusifiers also act as freezing-point depressants. The alkali metal salt of N-coco beta amino butyric acid when used in the amount of 2.5% based on latex solids is reported to be an effective freeze-thaw stabilizer. Polyalkylene polyamines,[6] polysubstituted phenates,[12] modified glyceryl monoricinoleate,[9] urea, and thiourea[4] are recommended to impart freeze resistance to emulsion paints.

Water-type paints will sometimes have bloom when used over plaster and other masonry surfaces. Barium compounds are reported to prevent bloom in emulsion paints.[2]

BUFFERS

pH is the logarithm of the reciprocal of the hydrogen ion concentration or really the hydrogen ion activity. This measure holds for aqueous systems only and has a scale from 0 (strongly acid) to 14 (strongly alkaline), with the neutral point at 7. A unit change in pH corresponds to a tenfold change in acidity.

The pH of a water-type paint has a profound effect on the physical properties and performance of the system. Buffers have the function of maintaining the pH level of the paint. Each latex system is more stable in a specific pH range. Surface agents, particularly those of the anionic type, have pH ranges in which they are the most effective.

Many protective colloids, such as carboxymethyl cellulose, casein, polyacrylates, and alginates, are soluble only in alkali and become insoluble at low pH. Particularly high pH will hydrolyze materials such as proteins, and so forth. Maintenance of a high pH in the package keeps certain bacteria which might cause spoilage in check. The use of permanent alkalis such as sodium hydroxide and other salts to adjust pH should be avoided, since water-soluble materials will be left in the film, thereby hurting water resistance. If possible, a volatile alkali such as ammonia should be used for adjusting the pH of the paint. Since ammonia is volatile, it can be lost from the batch during manufacture and cause difficulties such as thickening and finally coagulation of the paint.

The effect of pH on the viscosity of a linseed oil emulsion is shown in Figure 10.1 and one pigment dispersion in Figure 10.2.

COALESCING AGENTS

These are high-boiling solvents incorporated in emulsion formulas to aid in film formation. These solvents may be referred to as coalescing agents, film formers, or fluxing agents.

In most emulsion paints, the binder should be soft enough to coalesce at temperatures of as low as 40°F but hard enough to form a dry tack-free film as soon as the water is evaporated. Emulsion polymers are of a wide range of types. They can be soft and tacky, forming films at room temperature, or hard materials which must be softened before films can be produced.

A typical vinyl acetate homopolymer latex has a minimum filming temperature (MFT) of about 70°F. A 10% ethyl acrylate vinyl acetate copolymer

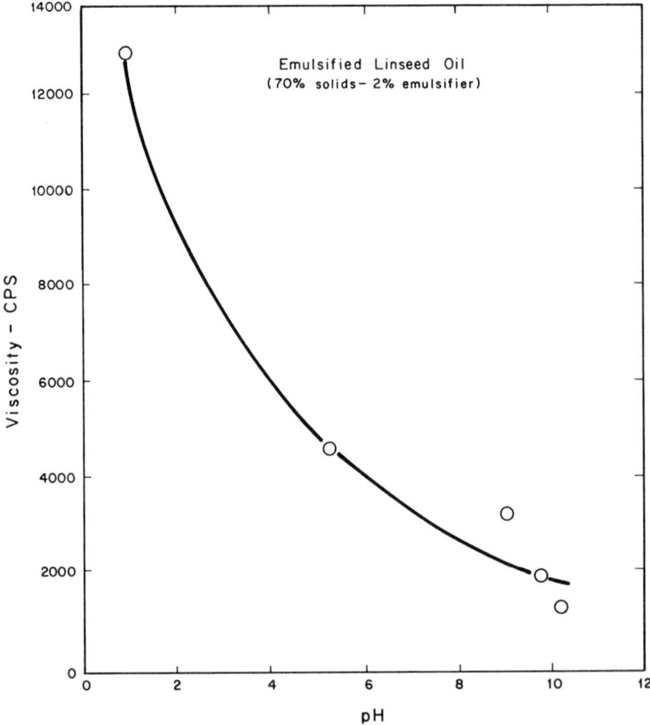

FIGURE 10.1. Effect of pH on viscosity

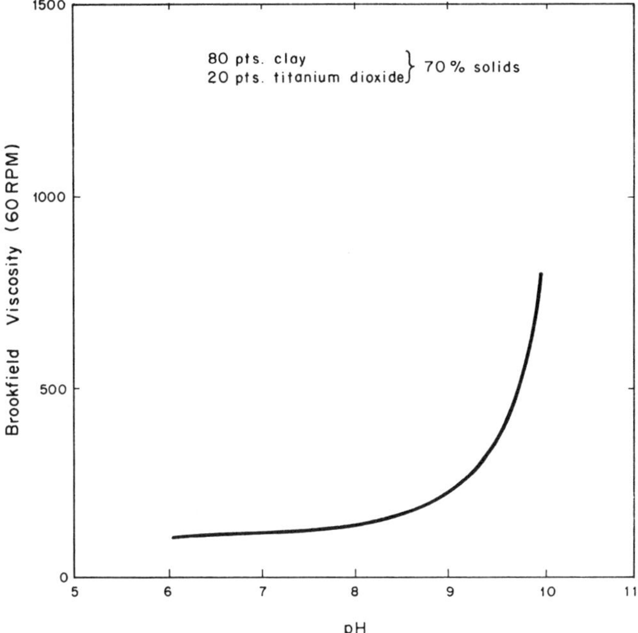

FIGURE 10.2. Viscosity vs. pH. Pigment dispersion.

emulsion resin will have a MFT of 60°F. Ten percent of dibutyl phthalate added to a polyvinyl acetate homopolymer will lower the MFT to 50°F.

The coalescing agent will improve leveling, scrub resistance, gloss, adhesion, and enamel holdout. However, package stability, drying, and recoatability may be affected adversely if excessive amounts are used. There are many materials that can be used as coalescing agents, but based on performance and odor the following are most commonly used:

Carbitol*	Butyl cellosolve acetate*
Carbitol acetate*	Butyl carbitol acetate*
Hexylene glycol	

The usual amount recommended for PVA paints is 10 to 40 pounds per 100 gallons of paint.[3,5]

A study was made of the use of various ether-alcohol materials in styrene-latex paint[11] at PVC's of 30 to 45%. The most effective material used was about 40 pounds of hexylene glycol per 100 gallons of paint. Leveling, flexibility, and scrubbability were all improved, especially at the higher PVC's.

Work with film-forming agents in acrylic emulsion paints showed that the addition of solvents such as hexylene glycol improved the durability.[1] Many

*Registered Trademark—Union Carbide Corp.

materials were evaluated for the effective filming of polyvinyl acetate latex paints.[10] Nitroalkanols are reported to be effective film formers.[7] The addition of polyols may slow the drying time to prolong the "wet edge" time of a brushing paint. Polyols will reduce "dry spray" in spray applications.

ANTI-RUST AGENTS

Waterborne paint use entails a problem not encountered with solvent paints. The water in the paint tends to promote corrosion of the metal container. As a rule alkaline paints will be less corrosive than those of a lower pH. Additives such as sodium benzoate added at a rate of 0.5 to 2.0% by weight will inhibit corrosion in PVA paints.[8]

Coating of a well-designed container with a protective coating is probably the best guard against corrosion. In certain coating operations where steel parts are coated with waterborne paints, a phenomenon known as flash rusting may occur. The surface may be mottled with rust spots before the coating is dried. Anti-rust agents will usually prevent this trouble.

DEFOAMERS

Foam can appear during the manufacture of the paint, slowing down production, and it can appear again at the time the paint is applied, causing cratering and breaks in the film. Foam can arise from the surfactants and thickeners present in emulsion formulas or from the amine resin combination in water-soluble paints. It may occur during the pigment paste milling and the letdown phases of manufacture. The best way to combat foam from mechanical causes is to eliminate faulty manufacturing procedures that allow air to be incorporated into the formulation. A case in point is the elimination of vortexing during the letdown phase.

The addition of from 2 to 4 pounds of a defoamer per 100 gallons of paint is usually sufficient. Excessive amounts can lead to "crawling" and "fish eyes" in the paint film. Defoamers can be added to either the pigment paste or the letdown.

In water-soluble systems defoaming is often a function of immiscible or borderline-miscible solvents. Odorless mineral spirits and octyl alcohol are classic defoamers, but are not the most effective. Esters frequently seem to be effective defoamers, among them butoxy ethyl acetate, butoxy ethoxy ethyl acetate, butyl acetate, and tributyl phosphate. In most cases a poor solvent for the resin and a water-immiscible liquid are generally more effective.

Antifoams and Defoamers

Foam is the dispersion of a relatively large volume of gas in a small volume of liquid. An *antifoam* is a surface-active agent that prevents foam formation.

An antifoam in one system may not be an antifoam in another. A *defoamer* is a surface-active agent that destroys foam after it has been formed. Because a successful defoamer spreads across the surface while breaking foam and prevents subsequent foam formation, the terms "antifoam" and "defoamer" are often used interchangeably.

An antifoam or defoamer consists of three major components:

1. Active antifoam components (destabilize foam)
   (a) Silicone oils
   (b) Hydrophobic silica
   (c) Organic polymers
2. Spreading agents (disperse the active agents)
   (a) Hydrophobic silica
3. Carriers (liquid medium for antifoam)
   (a) Hydrocarbon solvents

Foam in the paint can cause such problems as intercoat adhesion, floating, and flocculated aluminum pigment particles.

The following procedure may be used to screen defoamers:

Trade Sales Antifoam Evaluation

1. Prepare paint or varnish samples: 150 g are weighed into a ½-pint can.
2. Add antifoam: start with 0.5% of total sample weight.
3. Shake the can for ten minutes.
4. Check the weight per gallon.
5. Coat out.
6. Evaluate the wet and dried film for surface irregularities, color acceptance, and speed of bubble break.
7. Repeat for tint bases.
8. Repeat test after aging, usually at elevated oven temperatures.
9. Repeat the procedure for the more promising candidates to optimize dosage levels.

COSOLVENTS

Solvents used in water-soluble paints are called cosolvents or coupler solvents. They are usually about 20% of the volatile mixture, with the remainder being water.

Cosolvents help to solubilize the polymer, but more important they help to adjust the drying rate, improve the leveling and flow, control blistering during baking, and help in pigment dispersion. A list of cosolvents is shown in Table 10.1.

The flash point of the cosolvent determines the flash point of the paint. Viscosity behavior is modified with solvent; the substitution of solvent for water flattens the curve. (See Figure 10.3.)

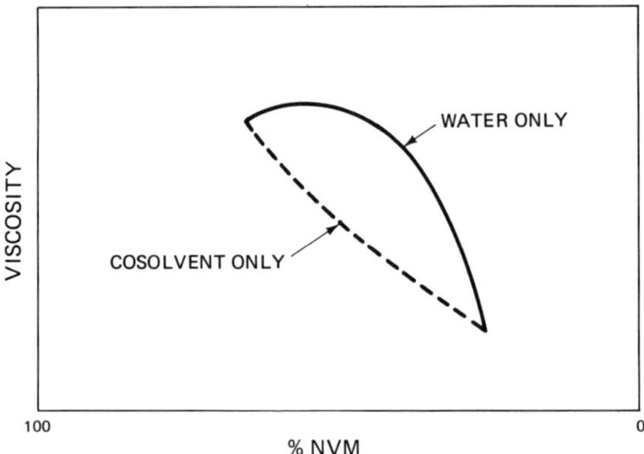

FIGURE 10.3. Reduction of an Alkyd in 80:20 Water:Cosolvent and 100% Neutralized with Ammonia.

Modification of evaporation rates is complicated by azeotropic or constant-boiling mixtures. Many of the common coupling solvents azeotrope in a mixture of about 80:20 water:solvent so that the solvent is gone from the coating at about the same time as water. Flow properties are frequently improved by adding butanol or propanol as secondary coupling solvents. Fast alcohols such as isopropanol may leave the coating before the water, thus aiding in sag control on vertical surfaces.

AMINE NEUTRALIZERS

Amines are widely used in waterborne coatings to increase resin solubility or reducibility and in the case of emulsions to improve stability. In the case of emulsion systems they are sometimes used as pigment dispersants. Amines adjust the pH of the coating system. The neutralizing amine may affect stability, film discoloration, drying rate, gloss, foam, and cost.

The amount of amine required is dependent upon the acid value of the resin, as shown by the following equation:

$$\text{weight amine} = \frac{(\text{mol. wt. amine}) \times (\text{acid value of resin solid}) \times (\text{wt. resin solids})}{56.1 \times 1000}$$

The amount of amine is often referred to as 100% neutralization or theoretical neutralization. Practical experience indicates that 80–90% of this amount is sufficient with many resin-amine combinatons. Dispersions may be made in the 40–60% neutralized range; however, as percent neutralization decreases, the appearance and behavior become more like those of an emulsion. Table 10.2 lists properties of some common amines.

Different resin systems require different pH ranges, as shown in Table 10.3.

TABLE 10.1. COUPLING SOLVENTS.

Coupler	Molecular weight	Boiling point, °C	Heat of vaporization, BTU/760 mm	Evaporation rate/butyl acetate = 100	Flash point, F	Surface tension, Dynes cm^2	Solubility parameter
Water	18	100	1074	16	—	72	23.5
Ethoxy ethanol	90	136	192	32	130	27	9.9
n-Butoxy ethanol	118	171	162	6	165	35	8.9
n-Butoxy diethylene glycol	162	230	130	< 0.01	240	30	8.9
Isopropyl alcohol	60	82	286	101	67	24	11.5
n-Butyl alcohol	74	118	141	45	97	24	11.4
Ethylene glycol	62	198	371	< 1	240	48	14.2

TABLE 10.2. PROPERTIES OF AMINES.

Compounds	Molecular weight	Boiling point, °C	Remarks
Ammonia	17	−33	Least expensive, fast volatilization, poor stability.
Monoethanolamine	61	170	Slow volatilization.
Monoisopropanolamine	75	159	—
2 Amino 2 methyl-1-propanol	89	165	High gloss with baked enamels.
Diethylamine	73	55	Fast volatilization, good package stability.
Morpholine	87	128	—
Diethanolamine	105	268	Slow volatilization, most foam.
Trimethylamine	59	−3	Fast volatilization.
Triethylamine	101	89	Less discoloration on baking, good package stability, least foam.
N,N-Dimethylethanolamine	89	134	Less discoloration on baking, good package stability, high foam.
Triethanolamine	149	335	—
N,N-Diethylethanolamine	117	162	Very good throwing power for electrodeposition.
N-Methyldiethanolamine	119	247	Less discoloration on baking.
N-Ethyldiethanolamine	133	253	—
2-Amino-2-methyl-1-propanol	89	None	High amine efficiency.
2-Dimethylamino-2-methyl-1-propanol	117	99	—

TABLE 10.3. PREFERRED pH RANGES OF RESIN SYSTEMS.

Maleinized oils	8.0–9.5
Oil-modified alkyd solutions	7.5–8.5
Oil-modified alkyd dispersion	6.5–7.5
Copolymer alkyd solutions	7.5–8.5
Polyester solutions	7.5–8.5
Acrylic solutions	7.5–9.5
Copolymer oil emulsions	8.0–9.0

Amine Efficiency

Amine efficiency is a measure of the amount of amine needed to attain a specific pH in a specified resin system.

Water-soluble resin systems are dependent upon molecular weight and acid value of the resin, cosolvent level, and amine efficiency.

Generally, the higher the molecular weight of the acid-functional resin, the better the film properties; the lower the acid value of the resin, the better the water resistance of the film; the lower the cosolvent level, the less the air pollution and lower the cost.

Molecular weight is directly related to both acid value and cosolvent level. As the molecular weight of the resin is raised, so also must be the acid value and the cosolvent level of the system be raised. However, acid value and cosolvent level are inversely related. As the cosolvent level is lowered, the acid value must be raised at the expense of optimum water resistance.

The use of a highly efficient amine can result in improvement of one or more of the other three systems parameters. As the amine efficiency goes up, the acid value goes down, the amount of cosolvent goes down, and the molecular weight of the resin can increase.

Three factors influence the efficiency of an amine in solubilizing an acid-functional resin. These factors are the molecular weight of the amine, the base strength of the amine, and the amine's solubility parameter. In general, the lower the molecular weight of the amine, the greater the efficiency, since each molecule contains one amine nitrogen. The base strength of the amine is related to the side chains attached to the amine nitrogen. The solubility parameter is dependent on molecular structure. In Table 10.4 amine efficiency factors are shown at a pH of 8.5.

DRIERS

Driers are used in waterborne paints that contain unsaturated polymers which cure by oxidation, such as oils, alkyds, oil-modified epoxies, oil-modified polyurethanes, and so on. Also styrene-butadiene latices which are partially unsaturated sometimes use driers in baking applications.

TABLE 10.4. AMINE EFFICIENCY FACTORS (AT 8.5 pH).

Amine	% Theoretical neutralization	Weight of amine to neutralize 100 g resin (AV 56.1)	Amine efficiency
2-Amino-2-methyl-1-propanol	110	9.73	1.00
Triethylamine	107	10.76	.90
2-Dimethylamino-2-methyl-1-propanol	105	12.28	.79
N,N-Dimethylethanolamine	143	12.69	.77
N,N-Diethylethanolamine	110	12.79	.76
Morpholine	167	14.49	.67

Driers for use in waterborne paints are available in a form readily dispersed in water. They are: 4% calcium, 6% cobalt, 6% manganese, 6% zirconium, and 24% lead.

Many water-soluble alkyd paints will dry without driers, but the addition of driers will improve humidity and salt-spray resistance.

The most economical drier combination for most systems is manganese with promoted cobalt:

0.06% Manganese
0.003% Cobalt phenanthroline chloride

Another recommended drier combination to reduce drying time is:

0.5% Active 8′[a]
0.1% Cobalt

For higher temperature (i.e., above 250°F) use rare earth driers and zirconium.

Calcium driers sometimes cause stability problems.

Zirconium has been found to be the best alternative to lead.

Loss of dry on aging of ambient drying waterborne alkyds is a problem.

To determine paint stability and loss of drier in a specific paint formulation, a batch of the paint is made up without driers and split into two parts:

(a) Test paint without drier
(b) Test paint with 0.2% cobalt as cobalt octoate on vehicle solids

Age all paints 24 hours at 77°F, and test for drying, viscosity, and so on. Age paints for five weeks at 77°F; then take one half of portion (a) and add 0.2% cobalt, and retest all three paints for drying, viscosity, and so on. If sample (a), to which no driers had been added, has deteriorated (i.e., shows

[a](35–40% 1, 10 phenanthroline) R. T. Vanderbilt Co., Inc.

any appreciable changes in viscosity, pH, or separation), or if (a) does not respond to added driers, the problem is one of unstable paint. The most common cause of instability is the reaction of the vehicle with water (hydrolysis). In this case, it is necessary to select a different vehicle, change the neutralization process, and so forth. If paint (b) is stable but has lost dry, replace the cobalt octoate with cobalt phenanthroline octoate at the same cobalt level.

REFERENCES

1. Burrell, R. W. S., *Paint Technol.,* **25,** No. 11, 15, (1961).
2. Broch, M. L. (to Firestone Tire and Rubber Co.), U.S. Patent 2,702,284 (1955).
3. Cogan, H. P., *Offic. Dig. Federation Soc. Paint Technol.,* **32,** 1216 (1960).
4. Gehring, H. T., (to Sherwin-Williams Co.), U.S. Patent 2,683,699 (1954).
5. Hovey, A. G., *Dig. Federation Soc. Paint Technol.,* **32,** 1176 (1960).
6. Johnson, P. H. (to Firestone Tire and Rubber Co.), U.S. Patent 2,802,799 (1957).
7. Johnson, R. L., *et al.* (to Dow Chemical Co.), U.S. Patent 2,898,317 (1959).
8. Kneeland, L. E., *Am. Paint J.,* **44,** No. 2, 21 (1959).
9. Miller, V. A. (to Firestone Tire and Rubber Co.), U.S. Patent 2,822,341 (1958).
10. Schwahn, C. O., and Sullivan, W. M., *Offic. Dig. Federation Paint Varnish Prod. Clubs,* **30,** 1122 (1958).
11. Tess, R. W., and Schmitz, *Offic. Dig. Federation Paint Varnish Prod. Clubs,* **29,** 1346 (1957).
12. Willis, V. M. (to Sherwin-Williams Co.), U.S. Patent 2,773,849 (1956).

11
Emulsion Formation

Resin emulsions can be prepared by two different methods: emulsion polymerization and the postemulsification of the resin. As the term implies, emulsion polymerization is actually two processes carried out in a single operation. It is the polymerization of the monomer to the polymer as well as the emulsification of the polymer. Postemulsification of the resin involves the emulsification of the resin which may have been obtained from natural sources or produced from synthetic materials through polymerization by addition or condensation reactions.

EMULSION POLYMERIZATION

Most commercial latices are manufactured by a technique known as emulsion polymerization. There are many references on this subject.[1-3] In this procedure, the monomer is dispersed in water in small droplets, and the polymerization takes place within these droplets. Because of good heat transfer through the low-viscosity water phase, the heat of polymerization can be removed easily, and polymerization proceeds rapidly. High-molecular-weight polymers are easily produced.

Latices in use today are derived from the free-radical-initiated addition polymerization of unsaturated monomers of vinyl or maleic type. These are compounds containing the —C═C— structures with various groups such as chlorine, acetate, benzene ring attached to the carbon groups. When polymerized, the polymers can be considered as a substituted polyethylene. For example, vinyl chloride polymerizes as follows:

```
H  Cl      Cl  H  Cl  H  Cl  H
C══C  →  —C — C— C — C— C — C—
H  H       H   H  H   H  H   H
```

The monomers commonly used in emulsion polymerization are shown in Table 11.1.

The contribution of each monomer to aqueous systems is discussed by Cantor.[4] Marketing of monomers for paint latices is discussed by Ball.[5]

Polymerization

The polymerization process begins with the initiator dissociating into a free radical. The free radical collides with a monomer molecule to form a more

TABLE 11.1. MONOMERS.

Monomer	Formula	Boiling Point, °C
Vinyl chloride	$CH_2{=}CHCl$	−13.4
Butadiene	$CH_2{=}CH{-}CH{=}CH_2$	−4.5
Vinylidene chloride	$CH_2{=}CCl_2$	31.7
Vinyl acetate	$CH_2{=}CHO{-}C({=}O){-}CH_3$	72.5
Acrylonitrile	$CH_2{=}CHCN$	77.3
Ethyl acrylate	$CH_2{=}CH{-}C({=}O){-}O{-}C_2H_5$	99.3
Methyl methacrylate	$CH_2{=}C(CH_3){-}C({=}O){-}O{-}CH_3$	100.6
Styrene	$C_6H_5{-}CH{=}CH_2$	145.1
Dibutyl maleate	$CH{-}C({=}O){-}O{-}C_4H_9$ ‖ $CH{-}C({=}O){-}O{-}C_4H_9$	281.0
Dibutyl fumarate	$C_4H_9{-}O{-}C({=}O){-}CH{=}CH{-}C({=}O){-}O{-}C_4H_9$	285.0
Dioctyl maleate	$CH{-}C({=}O){-}O{-}C_8H_{17}$ ‖ $CH{-}C({=}O){-}O{-}C_8H_{17}$	203.0 (5 min.)
Dioctyl fumarate	$C_8H_{17}{-}O{-}C({=}O){-}CH{=}CH{-}C({=}O){-}O{-}C_8H_{17}$	216.0 (5 min.)

complex free radical which continues to combine with additional monomer to form a long polymer chain.

The free radical is formed by the decomposition of an initiator such as benzoyl peroxide, hydrogen peroxide, or potassium persulfate.

Benzoyl peroxide C_6H_5—C(=O)—O:O—C(=O)—C_6H_5 → 2 C_6H_5—C(=O)—O:

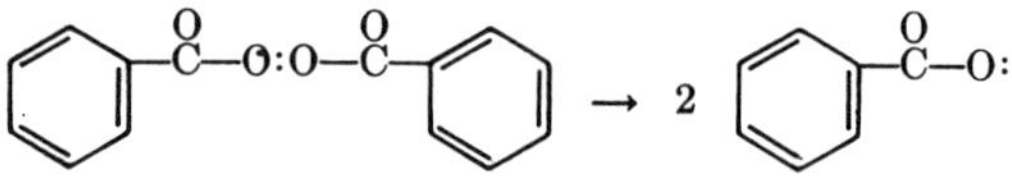

Potassium persulfate $S_2O_8^{=} \rightarrow 2SO_4$ Sulfate free radical

The second phase of the polymerization is known as propagation. The free radicals react with the monomer to open double bonds.

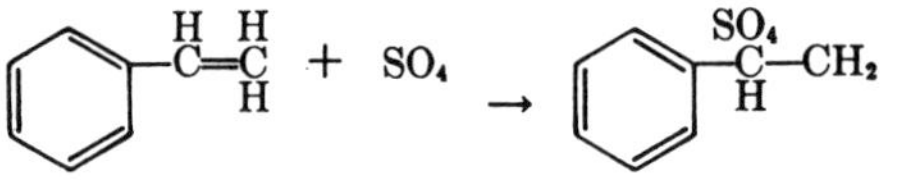

C_6H_5—CH=CH$_2$ + SO_4 → C_6H_5—C(H)(SO$_4$)—CH$_2$

Note that the radical from the initiator actually participates in the reaction. Therefore, one end of a polymer chain is a fragment of the initiator.

The final phase of the polymerization process is termination. This may occur in several ways, such as the combination of two growing chains, combination of a growing chain and a free radical, and so on.

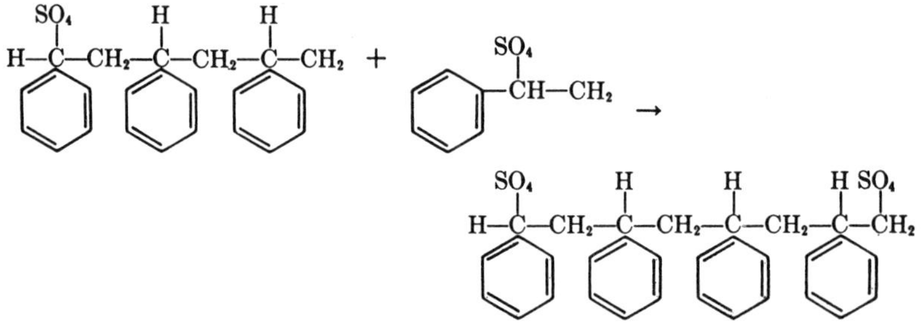

H—C(SO$_4$)(C_6H_5)—CH$_2$—C(H)(C_6H_5)—CH$_2$—C(H)(C_6H_5)—CH$_2$ + C_6H_5—CH(SO$_4$)—CH$_2$ →

H—C(SO$_4$)(C_6H_5)—CH$_2$—C(H)(C_6H_5)—CH$_2$—C(H)(C_6H_5)—CH$_2$—C(H)(C_6H_5)—CH$_2$—... (SO$_4$)

With monomers with a functionality of one (one double bond per molecule), linear polymers are formed. When a molecule with a functionality greater than one is incorporated into the system, branching and cross-linking may occur.

Linear

0—M—M—M—M— — — —M—M—T

Branched

0—M—M—M— — — — — —M—M—T
M
M
T

Crosslinked

0—M—M—M— — — — — —M—M—T
M
M M
0—M—M—M—M— — — — — — —M—M—T

0—initiator fragment
T—terminator fragment
M—monomer

Cross-linking can be accomplished by the incorporation of monomers with higher functionality, such as divinyl benzene, butadiene, and so on. In a solvent system, cross-linking would produce high viscosities and poor solubilities and might even produce gels or insoluble polymers. In a latex which is a dispersed system, the viscosity and solubility characteristics of the individual polymer particles do not affect the viscosity of the latex. In latices, a reasonable amount of cross-linking is desirable to toughen and insolubilize the final film.

Many of the commercial latices are produced using more than one monomer to attain the desired properties, such as hardness and elongation. For example, vinyl acetate by itself is a relatively hard, brittle polymer. It is copolymerized with dibutyl maleate to produce a softer more elastic polymer. This is known as internal plasticization. The softening point of polymer in a latex film is important, as it controls the fusion or coalescence of the latex particles. For example, a latex that has polymer particles with a softening point above room temperature will, upon evaporation of the water, produce a powdery deposit. If the softening point is close to room temperature, the film may fail to fuse in cold weather but will perform satisfactorily in warm weather.

Copolymerization

Copolymerization is any polymerization reaction in which two or more different monomers are converted into a homogeneous polymeric product. First, the composition of the copolymer depends upon the concentration of the monomers at the locus of reaction. Copolymerization is further complicated by the fact that monomers polymerize at different rates. Therefore, the composition of a copolymer will vary with the degree of conversion. Thus, the polymers formed at the start of the copolymerization may contain a higher proportion of the more reactive monomers than those formed at the end of the polymerization. A latex may contain a wide range of polymer molecules of varying composition and structures. The individual polymerization rates are also affected by temperature.

To prepare copolymers of more uniform composition, various procedures may be used. Sometimes the more reactive monomer is added slowly during the polymerization. Also temperature, pressure, or catalyst content may be varied during the polymerization.

One of the advantages of emulsion polymerization is the high-molecular-weight polymers that can be obtained by this method. Also, in an emulsion system, high-molecular-weight polymers can be handled readily in coating applications. In contrast, in solvent systems high-molecular-weight polymers require strong solvents and have excessive viscosities which make them difficult to handle. Increasing the molecular weight of the polymer usually increases toughness, scrub resistance, solvent resistance, alkali resistance, and so on.

The role of the surfactant in emulsion polymerization is discussed by McCoy,[6] particularly on its effect on rate of polymerization, particle size, stability, and molecular weight. Bobalek[7,8] reports on a dialysis technique for the study of soap transport phenomena in latices.

Theory

Emulsion polymerization is a complicated process. Harkins's publication[9] did much to explain the theory of emulsion polymerization. Prior to this time, it was thought that the polymerization took place within the droplet during emulsion polymerization, although this was not substantiated by the fact that the droplet is many hundreds of times bigger than the final latex polymer particle.

According to the Harkins theory, the monomer is distributed in three different phases in the emulsion system, i.e., dissolved to a small extent in the water phase, suspended in the form of monomer emulsion droplets, and solubilized in the surfactant micelles.

The process of initiation occurs in the aqueous phase, and the monomer radical diffuses into the micelles where polymerization occurs. The monomer emulsion droplets act as a reservoir for monomer. During the polymerization, the monomer emulsion droplets become smaller, and the micelles grow larger. It has been proven experimentally that the rate of polymerization is dependent on the number of particles present, which is largely dependent on the concentration of emulsifier. This is shown graphically in Figure 11.1.

The greater the amount of emulsifier present, the greater the number of micelles present and, finally, the greater the number of polymer particles, which are then smaller in size. (See Figure 11.2.) This is only an approximation and applies to one surfactant. Another surfactant might be more effective and would then produce a finer-particle-size latex at the same concentration.

Loebel[10] reports on the determination of average particle size of synthetic latices by turbidity measurements. The electron microscope is used as a quality control instrument for latex production.[11]

The first large-scale production of synthetic latices in this country was of GR-S rubber in World War II. It was developed by the U.S. Government Synthetic Rubber Program's agency "Rubber Reserve Company" as the "Mutual" recipe shown in Table 11.2. Following World War II there remained a tremendous industrial capacity for the production of GR-S emulsion polymers, but a greatly decreased demand for the product because supplies of natural latex were restored. As a result, other applications were sought for synthetic latexes. One of the most successful of these new applications was latex paints using styrene-butadiene latex with a higher styrene content.

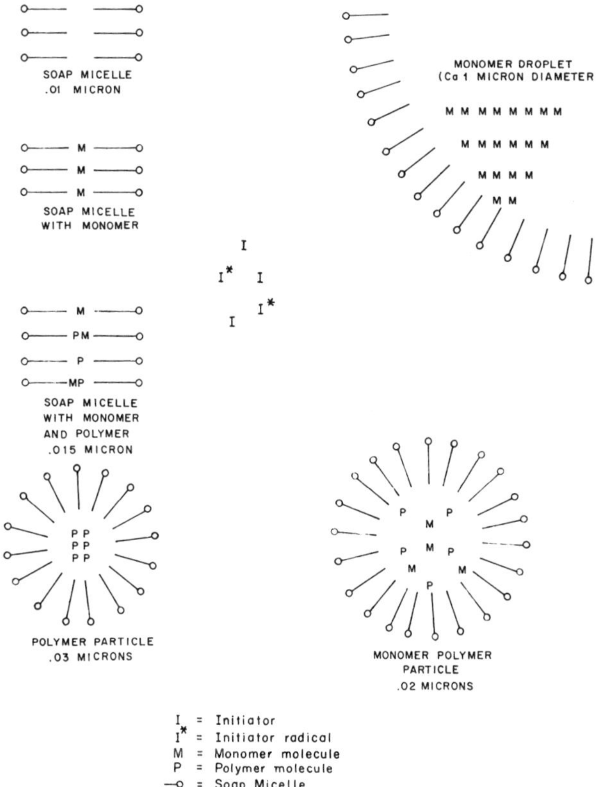

FIGURE 11.1. Emulsion polymerization concept.

Ingredients

The composition of a typical emulsion polymerization formula is shown in Table 11.3. Each ingredient will be discussed below.

Water

Water is the single largest ingredient in an emulsion polymerization reaction. For reproducible results, water of consistent purity must be used. This usually means distilled water or deionized water of about 250,000 ohms minimum resistance. Use of contaminated water may result in acceleration or inhibition of the initiator system, discoloration of the product, variation in particle size, or even coagulation of the latex during or after polymerization.

Surfactants

Choice of the polymerization surfactant is probably the most delicate and least understood feature of the design of emulsion polymers. This is especially true

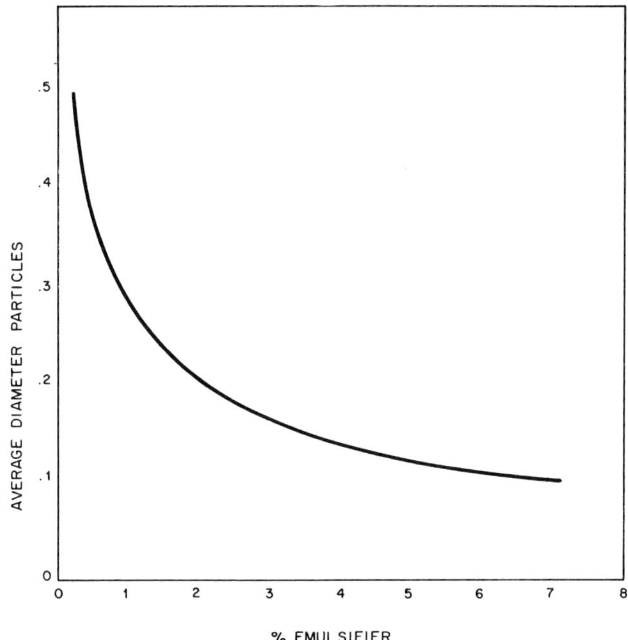

FIGURE 11.2. Emulsifier content vs. particle size.

for those products that are to be applied directly as the latex. The requirements for surfactant performance include micelle formation, solubilization of the monomer, stabilization of monomer droplets and the growing polymer particles, and stabilization of the final emulsion polymer. In addition, choice of surfactant must also be dictated by the desired properties of the final product—particle size, molecular weight, mechanical and freeze-thaw stability, foaming characteristics, water sensitivity, corrosion resistance, gloss, and many other characteristics. The relative importance of each of these factors

TABLE 11.2. "MUTUAL" RECIPE FOR GR-S SYNTHETIC RUBBER.

Component	Parts by weight
Butadiene	75
Styrene	25
Dodecyl mercaptan	0.5
Potassium persulfate	0.3
Soap flakes	5.0
Water	180.0
Polymerization temperature	50°C.
NVM	37%

TABLE 11.3. TYPICAL EMULSION POLYMERIZATION FORMULA.

Component	Wt. %
Water	50–60
Surfactants	1–5
Initiators	1–3
Monomers	40–50
Modifier	0–1
Modifier	0–1
Protective colloid	0–3

varies with the particular system of interest, and a balance must be reached. Unfortunately, there are few guidelines that can be followed in this regard, since structure–property relationships for surfactants in emulsion polymerization are virtually nonexistent. The situation is further complicated by the fact that the exact chemical composition of commercial surfactants is seldom known, and most such materials are complex mixtures of similar compounds. Considering the bewildering variety of products now available for use in emulsion polymerization, surfactant choice becomes very difficult. Patent literature, suppliers' literature, and experience help, but the most reliable solution to the problem is careful laboratory testing.

Both ionic and nonionic surfactants are commonly used in emulsion polymerizations. In general, cationic types are least desirable and are infrequently used. Anionic surfactants are probably the most widely used materials because they have low critical micelle concentrations and are very effective in reducing surface tension at low concentrations. Typical anionics include salts of C_{12}–C_{18} carboxylic acids (soaps), alkyl and/or aryl sulfates and sulfonates, ethoxylated sulfates and sulfonates, phosphates, succinates, and salts of polymeric acid derivatives. Potential problems associated with the use of anionics include foaming, sensitivity to electrolytes, and viscosity changes with time.

Nonionic surfactants are also often used in emulsion polymerization. These materials are usually ethoxylated alkyl alcohols or alkyd phenols having alkyl chain lengths of 8 or higher. The characteristics of these surfactants change widely with the degree of ethoxylation, the homologues (one to five ethylene oxide units) being oil-soluble and the more highly ethoxylated types being water-soluble. The major problem with the use of nonionics as sole emulsifiers for polymerizations is that they have lower surface activity than anionics and may not form micelles as easily. The result may be latexes of poor stability or with a grainy texture. This can be partially eliminated by using very highly ethoxylated types (30 or more ethylene oxide residues), which seem to provide stable polymerization sites even though they probably form hydrated aggregates rather than micelles.

A very common practice in emulsion polymerization is to use mixtures of anionic and nonionic surfactants to obtain a desired blend of properties. The advantages of this approach are several-fold and include particle size control, reduced foaming tendency, and improved stability.

Initiators

Free radical initiator systems in emulsion polymerization can be divided into two fundamental types: thermal and redox. Thermal initiators are compounds that undergo homolytic cleavage upon heating to yield free radicals. The most widely used thermal initiator in emulsion polymerization is the peroxydesulfate ion, added as the potassium, ammonia, or sodium salt. End-group analyses of polymers prepared using persulfate initiators show the presence of sulfate groups on the polymer, indicating that the major initiating species is the sulfate radical anion. In some cases, however, hydroxy groups are also found. This suggests that some sulfate radical anions undergo further decomposition before capture by micelles or particles. At temperatures below 40–50°C, the decomposition rate for persulfate is too low to be of practical utility in emulsion polymerization. Other thermal initiation systems sometimes include organic peroxides, inorganic perborates, percarbonates, and peracids.

Redox initiation systems are based on the fact that the rate of decomposition of highly oxidizing peroxy compounds can be accelerated by reducing agents. A very large number of such redox systems are known, and they all operate in very much the same way. In addition to the reducing agent, there is often also present a low concentration of metal ion that acts as an activator. A typical example of a redox system is persulfate, sodium formaldehyde sulfoxylate, and ferrous sulfate. Redox systems have been shown to be very effective in initiating emulsion polymerizations. One application of the technique is to allow preparation of polymers at much lower temperatures than is possible with thermal initiation. An example of this is the so-called cold SBR rubber, which is prepared at 5°C using redox methods as compared to 50°C for earlier "hot" recipes made using thermal initiation. Since the reaction rate is not decreased at the lower temperature, radical generation must be similar in spite of the large temperature difference.

A common practice in emulsion polymerization technology is preparation of latexes by redox technique in which part of the monomer charge is initiated at low temperature by redox and then allowed to exotherm. For example 30–50% of the monomer might be added, initiated at 10°C, and allowed to polymerize very rapidly with temperature rising to perhaps 80–90°C. The batch is then cooled and the procedure repeated with more monomer. Products prepared in this way may have smaller particle size and higher molecular weight than those prepared by thermal initiation techniques where monomer is added gradually over several hours at 70–80°C.

Monomers

Choice of monomers in emulsion polymerization reactions is largely determined by the end-use requirements of the polymer.

The first factor to be taken into account in the choice of a monomer or monomer mixture is the record order or glass transition temperature (*Tg*) of the polymer desired. This is the characteristic temperature at which the system undergoes a change from a hard, brittle material to a softer, more flexible one. Since polymers are generally unable to form films from latexes at temperatures below the *Tg*, an obvious requirement is that the polymer be above the *Tg* at the application and use temperature. A further restriction is placed on the *Tg* of the polymer by the fact that polymers become very soft at temperatures too far above the *Tg*, resulting in poor hardness, blocking, abrasion resistance, dirt collection, and so on. A balance must be obtained, therefore, between the good flexibility, adhesion, coalescence, and so forth, obtained at temperatures farther above the *Tg* and the good mechanical resistance properties found closer to the *Tg*. This is generally accomplished in coatings intended for ambient use by using polymers having *Tg*'s in the 0–30°C range.

Table 11.4 gives glass transition temperatures of homopolymers obtained from monomers typically used in emulsion polymerization. The equation at the bottom of the table enables calculations of approximate *Tg*'s for

TABLE 11.4. HOMOPOLYMER GLASS TRANSITION TEMPERATURES, °C.

	Tg		*Tg*
Acrylic acid	106	Methyl acrylate	9
Acrylonitrile	110	Methyl methacrylate	105
Butadiene	−85	*n*-Propyl acrylate	−45
n-Butyl acrylate	−56	*n*-Propyl methacrylate	35
n-Butyl methacrylate	22	Styrene	100
Ethyl acrylate	−22	*t*-Butyl acrylate	43
2-Ethyl-hexyl acrylate	−70	*t*-Butyl methacrylate	107
Ethyl methacrylate	65	*n*-Tetradecyl acrylate	20
2-Hydroxyethyl methacrylate	55	*n*-Tetradecyl methacrylate	−9
Isobornyl methacrylate	114	Vinyl acetate	30
Isobutyl acrylate	−40	Vinyl chloride	82
Isoprene	−73	Vinylidene chloride	−17
Isopropyl methacrylate	81		

$$\frac{1}{Tg} = \frac{W_1}{Tg_1} + \frac{W_2}{Tg_2} + \dots + \frac{W_n}{Tg_n}$$

This equation gives approximate multipolymer *Tg* in °K where W_n is the weight fraction of the monomer present and Tg_n is the homopolymer *Tg* in °K.

copolymers and multipolymers. Table 11.5 gives monomer compositions and *Tg*'s for polymers typical of those used in surface coatings.

Monomer choice is further determined by the chemical and physical properties of the final product. Some of the advantages and disadvantages of the general coating latexes are shown in Table 11.6. In addition, it should be noted that very often special monomers are included in a polymer preparation in order to impart specific properties. An example of this is the inclusion of hydroxy and carboxy functional monomers to give the final polymer reactivity toward cross-linking agents. Other examples are the inclusion of small amounts of acid for freeze-thaw stability, the use of difunctional monomers (e.g., divinyl benzene) to get cross-linking during polymerization, and the addition of ethylene to promote adhesion.

From the brief discussion above it should be clear that the choice of monomers is a very complex task. At all times polymer performance must be the primary criterion, but other factors such as cost, ease of production, long-term stability, and so forth, must also be considered. In addition, the ultimate relationship between the monomers and all the other formula components in an emulsion polymerization reaction may influence the choice. Finally, it should be pointed out that it is not sufficient simply to make intelligent choices of monomers, surfactants, and so on, because the gross emulsion properties, such as particle size, particle size distribution, surface tension, and so forth, also play a strong role in the performance of the latex.

Other Components

In addition to the major emulsion polymerization components discussed above, several other materials are sometimes used. These materials are buffers, protective colloids, and chain transfer agents.

TABLE 11.5. TYPICAL EMULSION SURFACE COATING POLYMER COMPOSITION AND *Tg* (WEIGHT PERCENT).

Acrylic		*Styrene Acrylic*	
Ethyl acrylate	65%	Butyl acrylate	55%
Methyl methacrylate	33%	Styrene	43%
Methacrylic acid	2%	Methacrylic acid	2%
Approximate *Tg* = 12°C		Approximate *Tg* = −5°C	
Styrene-Butadiene		*Vinyl-Acrylic*	
Styrene	60%	Vinyl acetate	85%
Butadiene	40%	2-Ethyl acrylate	15%
Approximate *Tg* = −6°C		Approximate *Tg* = 10°C	
Vinyl Acetate VVID			
Vinyl acetate	53%		
Vinyl versatate 10	47%		
Approximate *Tg* = 5°C			

TABLE 11.6. MONOMER CHOICE FOR EMULSION POLYMER SURFACE COATINGS.

Advantages	*Disadvantages*
Styrene-Butadiene	
Resistance to hydrolysis	Susceptible to oxidation
Solvent and chemical resistance	Residual odor
Low cost	Yellowing
Low specific gravity	High-pressure polymerization equipment required
Good on alkaline surfaces	
Suitable for metal finishes	
Acrylics	
Excellent clarity	Moderate to high cost
Stability to light and heat	Water resistance marginal
Outdoor durability good	Marginal adhesion (styrene-acrylics)
Gloss retention	Yellowing (styrene-acrylics)
Good pigment binding	
Small particle size	
Solvent resistance (acrylonitrile)	
Metal coatings (styrene-acrylics)	
Vinyls, Vinyl-Acrylics	
Stability to light	Susceptible to hydrolysis
Excellent adhesion	Poor solvent resistance
Low cost	Generally large particle size
Good exterior durability	
Grease resistance	

Buffers. Buffers are used to control the pH of the system before, during, or after polymerization. There are a variety of reasons why this may be desirable, including the following:

1. Some surfactant systems are very sensitive to pH, with respect to both the reformation of micelles and the stabilization of the polymer particles. Many latexes may coagulate if the pH is changed to a value outside the stability range of the emulsifier.

2. Initiator systems may be pH-dependent, with accelerated or retarded decomposition rates being observed under varying conditions, depending upon the exact formulation.

3. Hydrolysis of monomers and/or polymers may have to be considered. This is especially true of polyvinyl acetates at alkaline pH.

4. Copolymerization of some monomers may be better in certain pH ranges. For example, acrylic acid and methacrylic acid do not generally polymerize well above a pH of 5.

5. Monomer solubilities may be influenced by pH.

6. Product application requirements may establish a preferred pH range for

the latex. For example, a low-pH system would be undesirable for use on metal surfaces.

Protective Colloids. Protective colloids are used in some systems in place of surfactants or in addition to surfactants, to provide enhanced stability. Protective colloids are generally high-molecular-weight water-soluble materials such as polyvinyl alcohol, methyl cellulose, sodium algenate, hydroxyethyl cellulose, and so on. The surface activity of protective colloids may be lower than that of surfactants, but they are often better able to protect the surface of the growing polymer particles against agglomeration because they have lower mobility than surfactants. Protective colloids find their greatest utility in polymerization of polyvinyl acetate homopolymers and copolymers. In systems containing both protective colloids and emulsifiers, the distribution of the two species on the particle surface will depend on their relative adsorption potentials. Often surfactants are favored in this competition, and addition of a surfactant to a colloid-stabilized system may upset the absorption equilibrium, and flocculate the polymer. For this reason, there is usually a very delicate balance between colloid content and surfactant content in these systems.

Chain Transfer Agents. Chain transfer agents, which are sometimes called modifiers, are included in a polymerization formulation to regulate the molecular weight of the polymer. For latexes that are used as conventional surface coatings, it is usually desirable to maximize the molecular weight so that chain transfer agents are not used. However, for applications where the bulk polymer must readily flow, or where shorter chain length is advantageous for other reasons, a chain transfer agent may be added to the extent of a few percent. The most efficient chain transfer compounds are halogen and sulfur compounds, especially mercaptans. The most commonly used chain transfer agents are *n*-dodecyl or *t*-dodecyl mercaptans; however, they have strong odors and are hard to hide in a product latex.

Formulas

Formulas for typical coating latices—acrylics, styrene-butadiene, and vinyl acetate–acrylic—are shown in Tables 11.7–11.9. All the reactions are carried out in reactors constructed of inert materials, that is, glass or stainless steel equipped with agitators and condensers. It is usual practice to conduct emulsion polymerization reactions in an atmosphere of nitrogen to exclude traces of oxygen.

Acrylic Copolymer Latex. In Table 11.7 is shown a sample formula for the preparation of an acrylic copolymer using a two-stage redox initiation procedure.

Styrene-Butadiene Latex. In Table 11.8 is shown a sample formula for a styrene-butadiene latex using thermal persulfate initiation.

TABLE 11.7. SAMPLE FORMULA, ACRYLIC COPOLYMER.

Component	Parts by weight First stage	Second stage
Deionized water	1000.0 g	—
Alkylaryl polyether alcohol	31.6	35.0 g
Ethyl acrylate	233.0	283.0
Methyl methacrylate	168.0	188.0
Methacrylic acid	4.0	5.0
Ammonium persulfate	0.5	0.6
Sodium hydrosulfite	0.6	0.8

Procedure:

1. The components of stage one, except for the persulfate and hydrosulfite, are mixed well and cooled to 15°C.

2. The persulfate and hydrosulfite are added (separately), and the batch temperature reaches 65°C in 15 minutes. After 5 more minutes, the batch is cooled to 15–20°C.

3. The second-stage monomers and surfactants are added and mixed well.

4. The second-stage initiators are added, and the temperature reaches 65°C. This temperature is held for one hour.

5. The product is cooled to 300°C, strained, and the pH adjusted to 9.5 with ammonia.

TABLE 11.8. SAMPLE FORMULA, STYRENE-BUTADIENE LATEX.

Component	Parts by weight
Deionized water	125.0
Sodium dodecyldiphenyl ether disulfate	4.4
Dihexylsulfosuccinic acid soidum salt	4.4
Styrene	60.0
Butadiene	40.0
Potassium persulfate	1.3

Procedure:

1. A pressurized polymerization reactor is required.

2. The water and surfactants are added and mixed and cooled. The persulfate is added and dissolved.

3. Styrene is added, followed by butadiene using about a 3% excess for purging.

4. The reactor is sealed and the batch agitated for 16 hours at 60°C. The end of the reaction is indicated by a pressure drop due to the polymerization of butadiene.

TABLE 11.9. SAMPLE FORMULA, VINYL ACETATE–ACRYLIC LATEX.

Component	Parts by weight
Deionized water	75.0
Sodium bicarbonate	0.2
Potassium persulfate	0.3
Vinyl acetate	93.0
2-Ethylene acrylate	7.0
Ethyl oxide–propylene oxide block copolymer	5.0

Procedure:

1. The water, bicarbonate, and persulfate are mixed well and heated to 68°C.
2. The surfactant is dissolved in the monomer mixture and the resulting solution added to the water phase over 2½ hours at 68–72°C.
3. The temperature is held at 68°C for an additional 15 minutes to ensure complete conversion.
4. The latex is cooled and strained to remove traces of coagulum.

Vinyl-Acrylic Latex. In Table 11.9 is shown a sample formula for a vinyl-acrylic latex using delayed monomer addition and thermal persulfate initiation.

Latex Particle Size

Although the composition of a latex is very important in determining properties, the physical state of the emulsion is also of great significance. A property that has a large bearing on latex performance in surface coatings is particle size, which has at least some relationship to all of the following paint properties: gloss, flow, leveling, pigment beading, hiding power, coalescence, viscosity, mechanical and freeze-thaw stability, porosity, permeability, and all secondary properties what depend on these properties.

Since latex particle size can be controlled to a certain extent by the polymerization technique, it is worthwhile to review those factors that influence the final particle size of the latex. The particle size is, of course, a function of the number of particles initiated, since the more particles formed, the smaller the final particle size will be. The number of particles depends on:

1. The type and concentration of emulsifier. The dependence here is such that large-particle-size latexes can be made only at very low concentrations of emulsifier (e.g., below 0.5–1.0% based on monomer). At higher emulsifer concentrations, the particle size decreases only very slowly with increasing concentration.
2. The electrolyte concentration, which influences the critical micelle concentration of the surfactant.
3. The rate of radical generation and, therefore, the initiator rate.

4. The temperature, which affects both the critical micelle concentration and the initiation rate.
5. The type and intensity of agitation, which determines the degree of emulsification and the efficiency of mass transfer.
6. Other variables more difficult to define that cause batch-to-batch variations in the number of particles formed.

From the above it should be clear that small-particle-size latexes are favored from these polymerizations that have high initial surfactant concentrations and rapid initiator decomposition. Further, the type of surfactant used will be important as well as the nature of the monomers, since the first initiation occurs on solubilized monomer molecules. Further, the type of surfactant used will be important as will the nature of the monomers, since the first initiation occurs on solubilized monomer molecules.

Particle size distribution is also an important consideration in many latex applications. In general, narrow distributions are observed in those cases where the particle initiation period is short and where particle stability is high. Wider distributions are found where initiation of particles can occur over longer times, such as when fresh surfactant is being added continuously during the polymerization. Wider distributions are also found when growing-particle stability is low and particles flocculate to reduce total surface area. This is experienced when there is insufficient surfactant to stabilize the total available particle surfaces. Such a situation is especially pronounced in polyvinyl acetate latexes stabilized by hydrophilic protective colloids.

To produce a satisfactory latex for paint is a complex problem. Partly this is due to the fact that many desirable paint properties are contradictory to each other with regard to the overall system characteristics required to achieve the desired performance. (See Table 11.10.) Dependence of paint properties on polymer properties is shown in Table 11.11.

TABLE 11.10. CONTRASTING PERFORMANCE REQUIREMENTS FOR LATEX PAINTS.

Storage, freeze-thaw resistances	vs.	Easy coalescence
Substrate wetting (low surface tension)	vs.	Good leveling and low foaming (high surface tension)
Good flow	vs.	Good sag resistance
High stability	vs.	High water resistance
Stabilization of hydrophilic pigments	vs.	Stabilization of hydrophobic organic colorants
Fast drying rate	vs.	Good wet edge
Easy brushing	vs.	Good single-coat hiding
Good flexibility	vs.	Good hardness
Good adhesion	vs.	Good dirt, mildew, stain resistance
High gloss (small particle size)	vs.	Good leveling (large particle size)
High performance	vs.	Low cost

TABLE 11.11. PROPERTY CLASSIFICATION IN EMULSION SYSTEMS.

Basic reaction variables	Inherent latex properties	Product requirements
Raw materials	Molecular weight	Stability
Temperature	Molecular weight distribution	Clarity
Procedure	Particle size	Washability
Formulation	Particle size distribution	Durability
	Viscosity	Gloss
	Monomer sequence	Flexibility
	Tg	Solvent resistance
	Conversion	Corrosion resistance
	pH	Color retention
	Density	Coalescence
	Flow	Brushability
	Cross-linking	Drying time
	Surface properties	Porosity
		Pigment bending
		Adhesion
		Hardness
		Tensile strength

Emulsion Polymerization Trouble-Shooting Guide

The following outline describes possible sources of difficulty in emulsion polymerization, and their treatment.

1. Reaction fails to start or continue:
   - (a) Too much inhibitor — Use monomers with low level of inhibitor—be sure monomer is distilled or washed; increase amount or activity of initiator.
   - (b) Inhibition by oxygen — Exclude air; flush with nitrogen; reduce rate of agitation.
   - (c) Ionic inhibition — Control heavy metal content (Fe, Cu, Cr, Ag); remove rust particles from components.
   - (d) Destruction of initiator by emulsifier — Avoid cationic emulsifier with persulfate; use a persulfate, change emulsifier.
2. Reflux rate of monomer too high (temperature rise too high):

Problem	Remedy
(a) Inadequate removal of heat	Use more water; cool mixture before initiation; use external cooling during reaction; use two-stage process or gradual addition of monomer.
(b) Reaction too rapid	Reduce amount of initiator; use initiator with longer half-life.
3. Reaction rate too slow:	
(a) Improper initiator	Increase amount of initiator; use more active initiator; reduce amount or activity of reducing agent, reduce amount of ferrous salt.
(b) Improper emulsification	Change emulsifier, change concentration of emulsifier; emulsify monomers prior to reaction.
(c) Improper temperature	Increase temperature; add monomers more slowly to increase reflux temperature.
(d) Improper agitation	Change speed or design of agitator.
(e) Inhibitor content too high	Use monomers with low level of initiator.
4. Low conversion:	
(a) Exhaustion of initiator	Use initiator with longer half-life, add several charges of initiator; use two initiators of different half-lives, reduce amount of reducing agent.
(b) Unreacted monomer in emulsion	Strip volatile under reduced pressure.
(c) Reaction temperature too low	Raise final temperature to complete reaction.
(d) Formation of coagulum	Use conditions to reduce formation of coagulum.
5. Foaming during reaction:	
(a) Reaction rate too fast	Reduce temperatures, add defoaming agent.
(b) Improper emulsifier	Use nonionic emulsifier; use TAMOL N in emergency, quench with cold spray.
(c) Sparging rate too high	Reduce sparging rate.
6. Hydrolysis of monomers—shown by thickening or pH drift:	

(a) pH too high	Hold pH below 7.
(b) Monomer too sensitive to hydrolytic condition	Use ester with longer alkyl chain; substitute methacrylates for acrylates.
7. Formation of coagulum:	
(a) Unclean equipment	Cleanse with alkali or ammonia solution, rinse, steam with live steam; scrape walls to remove adherent solids.
(b) Improper agitation	Reduce rate of agitation if emulsion is breaking; increase rate of agitation if monomers are not dispersed.
(c) Improper emulsifier	Increase amount; use protective colloids, nonionic emulsifier; add more emulsifier at end of reaction.
(d) Improper dispersion of monomers	Add emulsified or dissolved monomers.
(e) Improper pH	Determine effect of pH on coagulum formation.
8. Reaction mixture gels (mechanical instability) during reaction or in storage:	
(a) Catalyst shock	Add initiator in solution; use slow or stepwise addition of initiator.
(b) Hydrolysis of monomer product	Reduce pH below 7.0.
(c) Too much acidic monomer	Reduce proportion of acidic monomer.
(d) Ionic coagulum	Reduce salt content.
(e) Cross-linking	Check purity of monomers; use chain transfer agent; reduce degree of conversion.
9. Two phases form during storage (creaming-sedimentation):	
(a) Inefficient emulsifier	Use more emulsifier; use two emulsifiers; add second emulsifier after reaction; change emulsifier.
(b) Ionic coagulum	Use nonionic emulsifier; change catalyst from persulfate to peroxide.
(c) Loss of water by evaporation	Use closed container.
(d) Oxidation degradation	Store under nitrogen.
(e) Bacterial degradation	Add biocide.

Problem	Remedy
10. Solids content low:	
(a) Mischarge of monomers	Check measuring devices, flush lines to ensure complete addition of monomers into reactor.
(b) Mischarge of water	Check measuring devices; check for leaks in condenser; check for moisture in monomers, sparging gas.
(c) Loss of monomers during reaction	Use cooler condenser water; reduce sparging rate; increase rate in condenser, control reaction temperature.
11. Emulsion too viscous:	
(a) Particle size too low	Use conditions to enlarge particles.
(b) Emulsion unstable	Use two emulsifiers; use nonionic emulsifier; reduce or avoid ionic components.
(c) Hydrolysis of monomer or product	Lower pH—use methacrylate instead of acrylate.
(d) Solids content too high	Dilute with water.
12. Drifting pH during storage:	
(a) Hydrolysis of polymer	See item 6.
(b) Ions in initiators	Reduce amount of persulfates or bisulfites.
13. Discoloration of emulsion:	
(a) Inhibitor	Use lower level of inhibitor; remove inhibitor by washing or distillation.
(b) Contamination	Check components for impurities; check for corrosion of equipment, lines, etc.; filter components to remove rust particles.
14. Molecular weight too large:	
(a) Chain formation too few	Use more initiator; use initiator with shorter half-life; use higher temperature; increase rate of agitation.
(b) Chain termination too few	Use chain transfer agent.
15. Molecular weight too low:	
(a) Chains form too often	Reduce amount of initiator; reduce rate of addition of initiator with longer half-life; reduce temperature of reaction; reduce rate of agitation.
16. Particle size too large (chalky emulsion):	

Problem	Remedy
(a) Inefficient emulsification	Increase amount of emulsifier; pre-emulsify monomers; use anion emulsifier.
(b) Improper initiator	Use water-soluble initiator.
(c) Reaction temperature too high	Reduce temperature; use a redox process.
(d) Ionic coagulation	Use nonionic initiator; use nonionic emulsifier; check contamination in water and components.
17. Particle size too small (translucent emulsion):	
(a) Emulsifier content too high	Reduce amount of emulsifier; reduce proportion of anionic emulsifier.
(b) Improper initiator	Use less water-soluble initiator.
(c) Salt content too small	Add small amounts (0.05%) of NaCl.
(d) Monomer too soluble	Add alcohol, removable after reaction.
18. Emulsion mechanically unstable:	
(a) Agitation or pumping causes gelation	Reduce rate of agitation; change design of agitator; change pump.
(b) Insufficient emulsifier	Use nonionic emulsifier; add protective colloid.
(c) Uncontrolled pH	Use acidic comonomer (methacrylic acid); check effect of pH on mechanical stability.
19. Emulsion unstable in freeze-thaw cycle:	
(a) Copolymer too soft	Use copolymer formulation with increased hardness; use acidic comonomer (methacrylic acid).
(b) Coalescense during freezing of H_2O	Add antifreeze (glycols, alcohols); add humectant; use more nonionic emulsifier or protective colloid. Maintain low pH.

POSTEMULSIFICATION

Oil, resins, waxes, and so on, can be prepared in emulsion form by combining with a surfactant and water under proper mixing procedures.

For emulsification under normal procedure, the resin, oil, or wax must be in a liquid state or converted to a liquid state by the use of heat or the addition of a solvent.

There are several techniques for preparing emulsions in which the resin and water are combined by various mechanical means such as stirrers, mixers, pumps, colloid mills, and so forth. These techniques are:

1. *Agent in Water*. The surfactant is added to the water, and then oil is added slowly with agitation. This produces an O/W emulsion directly.

2. *Agent in Oil*. The emulsifying agent is dissolved in the oil phase. This mixture is added directly to water, forming an O/W emulsion. When the water is added to the oil-surfactant mixture, a W/O emulsion is formed first, which inverts to an O/W emulsion. This method produces an exceedingly fine-particle-size emulsion. Viscosity is at its maximum value at the inversion point. (See Figure 11.3.)

3. *Combination*. When a combination of surfactants is used, they may be put into both phases. The lypophilic emulsifier (HLB < 13) is put into the oil phase, and the hydrophilic emulsifier is put into the water phase.

4. *In situ*. In emulsions, either O/W or W/O, the fatty acid may be dissolved in the oil and the alkali in the water. The soap then forms at the interface.

5. *Alternate Addition*. In this method the oil and water are added alternately to the surfactant in small portions.

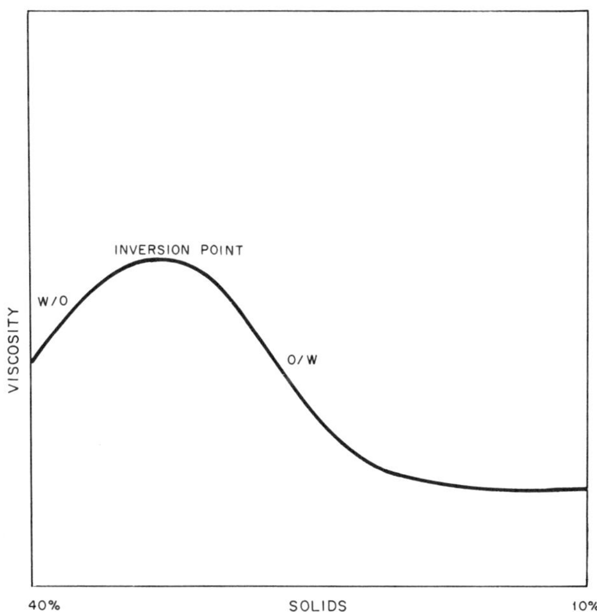

FIGURE 11.3. Inversion of an emulsion.

Solvents are required to prepare emulsions of many resins, such as nitrocellulose, ethyl cellulose, short oil alkyds, and chlorinated rubber. Since solvent is present in the oil phase, this increases the volume of the oil phase and, therefore, reduces the solids content of overall film former in the emulsion.

A simple formula for a mineral oil emulsion is as follows:

	Parts
Mineral oil	88
Triethanolamine	4
Oleic acid	8

Add the oleic acid to the oil and mix until clear. Add triethanolamine with mixing. Add the mixture to an equal amount of water with stirring.

The following example illustrates the preparation of an emulsion using an alkyd:

	lb
65% Linseed oil alkyd	100
Casein	8
Ammonia 29%	5
Water	100
Emulsifying agent	1
Sodium phenyl phenate	1
	215

The casein is dissolved in the water plus ammonia, together with the emulsifying agent and preservative, and heated to 80°C. The alkyd is heated to 80°C and poured in slowly with high-speed agitation.

STABILITY OF EMULSIONS

Emulsions produced must have good stability. The stability depends upon: the particle size; the difference in density of the two phases; the viscosity of the continuous phase; the changes on the particles; the nature, effectiveness, and amount of emulsifier used; the conditions of storage, including high and low temperatures, agitation, and vibration; and dilutions or evaporation.

Since the particles of an emulsion are freely suspended in a liquid, they obey Stokes' law unless they are charged. The rising and settling of the particles are termed creaming and sedimentation, respectively. The phenomenon of creaming receives its name from the most common example, the separation of cream in unhomogenized milk. This is not a break but rather a separation into two emulsions, one of which is richer in the disperse phase. When the richer phase goes to the bottom, the phenomenon is called "downward creaming." When creaming or sedimentation occurs, coalescence is

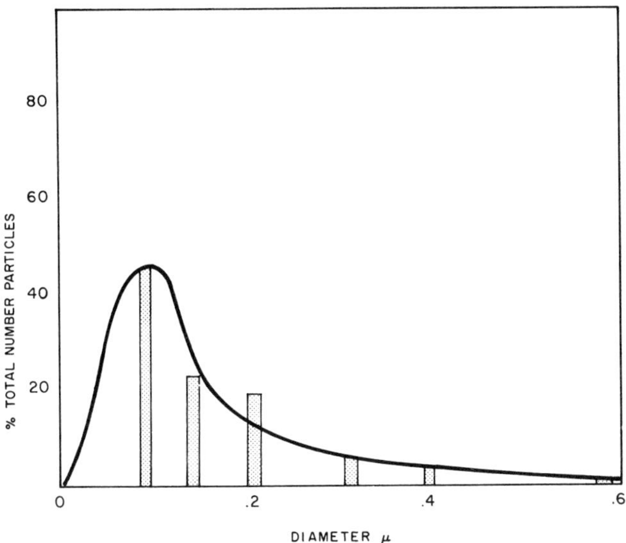

FIGURE 11.4. Particle size distribution in a typical latex.

more likely to occur because of crowding or physical contact of the particles.

Particle size data on emulsions usually give the average particle size. There is actually a range of particle sizes in any one emulsion. A typical particle size distribution curve is shown in Figure 11.4.

PHASE VOLUME

An emulsion consists of many spheres (internal phase) packed into a definite volume of space (external phase). According to solid geometry, an assembly of spheres of equal radius can be placed in a position of densest packing in two ways. In either case, however, the spheres (internal phase) occupy 74% of the total volume. Any attempt to exceed this volume will cause breaking or inversion.

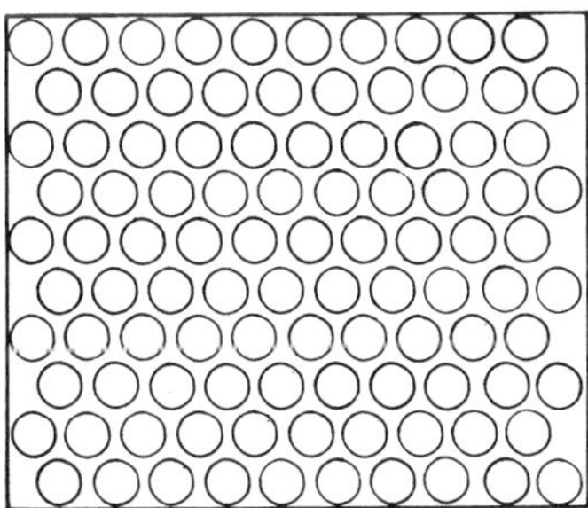

FIGURE 11.5. Maximum packing of spherical droplets is 74% of volume.

REFERENCES

1. Bovey, Kolthoff, Medalia, Meeham, *High Polymers,* **9,** (1955).
2. Alexander, A. E., *J. Oil Colour Chemists' Assoc.,* **45,** No. 1, 12 (1962).
3. Naidus, H., *Offic. Dig. Federation Soc. Paint Technol.,* **33,** 1582 (1961).
4. Cantor, H. A., *et al., Paint Ind. Mag.,* **75,** No. 8, 7 (1960).
5. Ball, E. J., *Paint Ind. Mag.,* **76,** No. 8, 22–3 (1961).
6. McCoy, C. E., *Offic. Dig. Federation Soc. Paint Technol.,* **35,** No. 459, 327 (1963).
7. Bell, E. G. and Bobalek, E. G., *Offic. Dig. Federation Soc. Paint Technol.,* **32,** 1047 (1960).
8. Bobalek, E. G., *Offic. Dig. Federation Soc. Paint Technol.,* **35,** 423 (1963).
9. Harkins, W. D., *J. Am. Chem. Soc.,* **69,** 1428 (1947).
10. Loebel, A. B., *Offic. Dig. Federation Paint Varnish Prod. Clubs,* **31,** 200 (1959).
11. Palen, V. W., *Paint Varnish Prod.,* **48,** No. 1, 38 (1958).
12. Loranger, A. H., Serafini, T. T., Fisher, W., and Bobalek, E. G., *Offic. Dig. Federation Paint Varnish Prod. Clubs,* **31,** 428 (1959).
13. Serafini, T. T., and Bobalek, E. G., *Offic. Dig. Federation Soc. Paint Technol.,* **32,** 1259 (1960).
14. Duck, E. W., "Emulsion Polymerization" in *Encyclopedia of Polymer Science and Technology.* H. F. Mark, Editor. Vol 5. John Wiley, New York, 1966.

12
Manufacture and Handling

The manufacture of waterborne products involves the handling of materials containing water and, in the case of emulsions, colloidal dispersions. Since water will rust iron and steel equipment, contaminating the batch with iron, corrosion-resistant materials must be used for much of the equipment. In the low pH ranges, water is more corrosive than at higher pH ranges. The ideal construction material is stainless steel, but it is not used for all installations because of the high cost. A wide range of materials, such as wood, glass-coated steel, stainless clad steel, resin-coated steel, and plastic, may be used.

A discussion of equipment should cover the manufacture of emulsions or latices, storage of the latex, the manufacture of the paints, and handling of the finished paint.

EQUIPMENT FOR POSTEMULSIFICATION

For the preparation of emulsion systems directly from a resin or oil, the following equipment may be used: mixers, homogenizers, colloid mills, and ultrasonic devices.

A common type of mixing operation is one or more propellers mounted on a common shaft in a mixing tank. Variations include location of propeller shaft, the use of two or more propeller shafts, and the use of complex propellers. Propeller agitation is most satisfactory for low- and medium-viscosity emulsions.

Turbine agitation includes fixed baffles either on the tank wall or adjacent to the propellers. Turbine-type systems give a very high degree of shearing action. Turbines may be used with high-viscosity materials.

The colloid mill consists of a rotor and stator with clearance of a few thousandths of an inch. With such small clearances, an extremely high shearing action occurs. Temperature rise is great because of the great amount of shearing action, and in most cases external cooling must be employed. Rotor speeds are from 1,000 to 20,000 rpm. Rate of output varies inversely with viscosity.

A homogenizer is a device in which dispersion is effected by forcing the mixture to be emulsified through a small orifice under very high pressure. A production homogenizer consists of a pump which provides the required pressure of 1000 to 5000 psi and a spring-loaded valve which constitutes the orifice.

Pebble and ball mills can be used to make emulsions and dispersions and are relatively slow-speed equipment. A more recent development is the use of high frequency or ultrasonic oscillators to prepare emulsions. This method seems to be best for low-viscosity liquids.

See Table 12.1 for data on emulsions prepared with various equipment.

EMULSION POLYMERIZATION[1]

Polymerization in the plant may be carried out in a steam-jacketed stainless steel kettle equipped with a turbine-type agitator, a vapor pipe leading to a multiple-tube condenser from which distillate can be drained off or returned to the reaction kettle (the condenser is not required when a redox system is used), a recording thermometer, a U tube containing fluid to indicate pressure inside the kettle, and a sight glass. The monomer emulsion is made up in an overhead stainless steel tank equipped with a turbine-type agitator and a tankometer. The emulsion is added to the reaction kettle by gravity through a hand-operated valve.

Agitation must serve two purposes: to provide sufficient shear to disperse adequately monomer droplets in the aqueous phase and to provide sufficient movement of the liquid mass to ensure good heat transfer to the jacket. The heat of polymerization for vinyl acetate is 440 Btu per pound, so that a temperature rise of 300°F may take place in a typical reaction. About 3 to 5 hp per 1000-gallon reactor should be provided. For high-viscosity emulsions, a minimum of 10 hp should be provided.

Low-speed, positive-action rotary or reciprocating pumps are satisfactory for handling mechanically stable finished emulsions. Pumps with close-fitting sliding valves should be avoided. Mechanically unstable emulsion cannot be handled in pumps with close tolerances, and diaphragm pumps are required.

Unreacted monomers must be stripped from the latex. This can be accom-

TABLE 12.1. PARTICLE SIZE RANGE (MICRONS).

	hp per 100 gal/hr	1% Emulsifier	5% Emulsifier	10% Emulsifier
Propellor	up to 2	no emulsion formed	3–8	2–5
Turbine	up to 4	2–9	2–4	2–4
Colloid mill	5–20	6–9	4–7	3–5
Homogenizer	3–5	1–3	1–3	1–3

plished by steam stripping, vacuum stripping, air or gas blowing, film evaporation, or spray stripping.

Filtering emulsions is difficult. They are usually strained through a 100-mesh screen, which must be washed repeatedly during the operation.

While most latices are produced using a batch process, they may be produced on a continuous process.[5] The materials pass through three or more continuous polymerization reactors. The continuous process results in a greater uniformity of product.

STORAGE OF LATEX

The preferred materials for latex storage tanks are stainless steel of types 304, 316, or 347. A stainless steel tank can be cleaned easily without damage. The preferred shape is a vertical cylinder with as little surface area as possible. A skin will form on the surface of the latex in storage; and, as the liquid level moves up and down, a film of resin accumulates on the wall of the tank. The top of the tank should have a manhole for inspection and access for cleaning. Storage tanks are usually 3,000- to 10,000-gallon capacity. Cost averages 60¢ per gallon capacity. The tank should be fitted with at least a 2-inch line for filling, which should extend to about one foot off the bottom. The tank should be fitted with at least a 3-inch line for emptying. The tank should be fitted with a 1-inch vent pipe. Tank temperature should not go below 40°F nor exceed 110°F, so that it is preferable to have tanks located within a heated building. (See Figure 12.1.)

Piping and Valves. The minimum-size lines recommended for handling latex are 2 inches and preferably of stainless steel. The line emptying into the paint tank should be fitted with a screen to remove any skins. Valves can be of stainless steel or bronze with stainless steel trim.

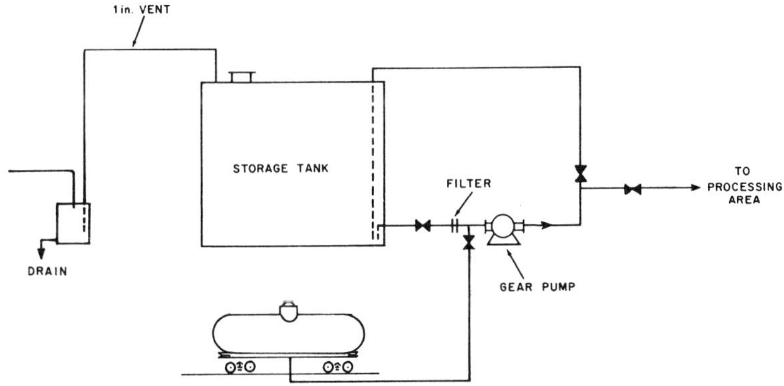

FIGURE 12.1. Latex storage system.

LATEX PAINT MANUFACTURE[2,3,4]

In making emulsion paints the protective colloid, dispersant, inert, and pigments are mixed with enough water to produce a paste consistency. Mixers are slow-speed with large blades which produce high shear and scrape the walls of the tank to mix all dry stock. After a satisfactory mix is obtained, paste is run through a mill such as the Morehouse mill and then into the thinning tank. The mix tank should be in the top level and fed by gravity to the thindown tank. Here the latex is added and the batch shaded and adjusted to standard characteristics. Mixing in this tank is slow, and shear is at a minimum. Wash equipment with water and save the wash water for use in the next batch.

WATER-SOLUBLE PAINT MANUFACTURE

In making water-soluble paints the pigments and inerts are dispersed with a portion of the resin and then transferred to the thindown tank where driers and the remainder of the resin are added. Wash equipment with plain water and/or 10–20% butyl cellosolve in water solution if it is cleaned immediately after use. If paint has started to dry, add some ammonia to the water solution.

DISPERSING EQUIPMENT OR MILLS

Since some pigments are harder to disperse than others, different types of dispersing equipment or mills are used. The dispersing of pigments in a liquid is often called grinding although this is a misnomer, as there is very little reduction in size of the original pigment particles during the dispersion operation in paint manufacture.

High-Speed Stone Mills

The high-speed stone mill consists of a stationary carborundum stone and a high-speed rotating stone. Pigment pastes are passed between these stones, and the distance between the stones can be varied for more or less shearing action. These mills are suitable for high production rates of fairly easy to disperse pigments for architectural paints where very fine dispersion is not required.

Roller Mills

Roller mills consist of steel rollers that rotate in opposite directions at different speeds. The pigment–liquid paste is passed between the rolls, which can be adjusted to different clearances. These roll mills are the most widely used in the paint industry. Roller mills have relatively slow production rates and require skilled operators, but are capable of producing fine dispersions. They are usually not used for waterborne systems except when the pigment is ground in a solvent system without water to produce a water in oil, without emulsion.

Dough Mixers

Heavy-duty dough mixers, consisting of two roughly S-shaped blades that overlap and rotate in opposite directions, are sometimes used to disperse very heavy pastes.

High-Speed Disperser

A high-speed disperser such as the Cowles consists of a tank containing a circular impeller driven at high speed by a vertical shaft. Dispersion of the pigment–liquid mixture is achieved by high shear action developed near the surface of the impeller. High-speed dispersers are used where very fine dispersion is not required, or whenever the pigments will disperse easily in the liquid. The production rate is very high, and this type of equipment is used to manufacture most architectural paints. Temperature should be controlled to a maximum of 140°F to minimize tendency to foam. The preferred viscosity is in the range of 75–90 K.U. used for pigments such as titanium dioxide, iron oxides, and so on.

Ball and Pebble Mills

Ball and pebble mills consist of large cylindrical steel tanks that rotate around a horizontal axis. The mill is partially filled with steel or procelain balls or pebbles and the material to be dispersed. Baffle bars are usually added to the side of the tank to help lift the balls for better dispersion. Steel balls are the most efficient, but cannot be used to produce white paints. They are not recommended for waterborne paints. Pebble mills require very little attention after they have been charged and are capable of producing good dispersions. Temperature must be controlled, or excessive foaming will occur.

Sand and Bead Mills

The sand mill consists of a cylinder containing coarse sand as a grinding medium. The pigment paste to be dispersed is fed into the mill, and rotating impeller discs driven by a vertical shaft impart a circulation pattern to the sand paste mixture. There is a difference in velocity between the particles near the surface of the impellers and the rest of the material, which develops a high shear action to disperse pigments. Pigment slurries can be passed through these mills for continuous operation. Production rates of sand mills can be fairly high, and dispersion is quite good. These mills are often used for high-quality industrial finishes. They are used for dispersing pigments such as Hansa yellow, phthalocyanine greens and blues, and so forth.

PIGMENT SLURRIES

Slurry is shipped either in tank trucks or rail cars. At the plant unloading facility a pump or sufficient air pressure is needed for unloading. Rail cars containing 10,000 gallons can be emptied in about 2 hours. Stainless steel and hard-rubber-lined pumps as well as pumps of silicon iron and other abrasion-resistant alloys can be used.

Slurry storage tanks can be constructed of stainless steel, reinforced plastic, or rubber-lined or epoxy-lined mild steel. Tanks should be large enough to hold a reasonable inventory—about 15,000 gallons for rail cars and 5,000 gallons for truck deliveries.

A low-speed agitator equipped with a timing device for intermittent agitation is sufficient to prevent settling. Each slurry grade requires its own storage tank and distribution line. Outdoor tanks should be insulated or heated if necessary to guard against freezing in extremely cold weather. Storage tanks should be sealed and provided with a small vent. Humidity should be kept relatively high in the tank to prevent the formation of dried pigment on the walls.

A recirculating loop can be used to distribute the slurry to use points within the plant. Distribution lines, like storage tanks, should be stainless steel or reinforced plastic. A magnetic flowmeter is used to dispense slurry into the dispersers or mix tanks.

REFERENCES

1. Baum, S. J., *Ind. Eng. Chem.*, **49,** No. 11, 1797 (1957).
2. Baumhart, E. E., *Offic. Dig. Federation Soc. Paint Technol.*, **32,** 662 (1960).
3. Bowden, A. P., *Paint Manuf.*, **30,** No. 7, 241 (1960).
4. Peale, S., *Paint Ind. Mag.*, **75,** No. 3, 18 (1960).
5. Toomey, R. F., *Paint Ind. Mag.*, **76,** No. 1, 16 (1962).

13
Trade Sales Paints

Trade sales paints are shelf good paints for use by the do-it-yourself painter and the professional painter.

They are available in waterborne form as:

Wall primers and sealers—clear and pigmented
Interior wall finishes—flat and semigloss
Interior trim finishes
Wood stains
Multicolored coatings
Tilelike glaze coatings
Concrete floor and patio paints
Exterior house and trim paints—flats and semigloss
Exterior shingle stains
Stucco and masonry paints

Trade sales paints that are not now generally available as waterborne coatings are:

Varnishes, lacquers, shellacs—interior or exterior
High-gloss enamels
Swimming pool paints

The popularity of waterborne paints, particularly with the homeowner who does his own painting, is attributable to ease of application, fast drying characteristics, ease of cleanup with soap and water, and lack of solvent odors. Also, the paint may be applied to a damp surface without adverse effects and may be recoated the same day. Trade sales or consumer waterborne paints are usually of the latex type.

Trade sales paints used by the professional painter may be the same as those of the do-it-yourself painter; or, in many cases, they are products formulated for spray application. They should have one-coat coverage.

The first latex paints used by the consumer were flat interior wall finishes. Then exterior masonry paints were developed. Exterior housepaints came next, followed by semigloss interior and exterior products.

The first binders used were styrene-butadiene latices for interior finishes.

While styrene-butadiene paints have good alkali resistance, they tend to oxidize further on aging, becoming yellow and brittle. Styrene-butadiene latices have now been almost completely replaced by other latices.

Polyvinyl acetate latices are used in both interior and exterior paints. They are lower in cost than acrylics and have good color retention and grease and oil resistance. Acrylic latices are higher in cost, have good color retention and durability, and develop good water resistance soon after application.

A trade sales line of paints must be formulated to accept universal color-tinting systems in order to make available to the consumer a wide range of colors.

The classification of paints according to gloss is shown in Table 13.1. Measurements are made at a 60° angle.

The latest trend in trade sales paints is to low-gloss "egg-shell" interiors and high-sheen exterior latex paints. These paints look flat when viewed head on, but have a sheen when viewed from an angle. This type of finish retains the "new look" for a long period of time. Having a smoother, more impervious surface then flat paints, they are more dirt-resistant and easier to clean.

CASEIN PAINTS

Casein paints are applied to masonry and plaster surfaces.

Casein paints usually contain about 10% casein with some lime to insolubilize the casein after it is applied. Casein paints shrink on drying, and may pull away from smooth surfaces.

A typical formula is as follows:

Whiting	50 lb
Clay	20 lb
Dextrin	2 lb
Casein	12 lb
Lime	15 lb
Trisodium phosphate	1 lb
Bichloride of mercury	1 oz

TABLE 13.1. CLASSIFICATION OF PAINTS ACCORDING TO GLOSS RANGE/ PIGMENT VOLUME CONTENT.

Paint	Gloss, % (@60°)	Pigment volume content, %
High gloss	70–95	10–15
Semigloss	30–70	15–30
Egg shell—interior and exterior	10–25	30–35
Flat—interior and exterior	2–10	35–60

Dry blend all materials. For application use 10 pounds of powder per gallon of water.

PRIMERS—SEALERS

Latex paints are particularly well suited for use on porous surfaces such as masonry and plaster. Latex particles, although very small, are still large enough so that they will not penetrate into a porous surface in the same manner as a polymer in solution, and thus are excellent sealers because they have outstanding "holdout."

Large particle size (i.e., 0.3- to 1.0-micron polyvinyl acetate latices) have been used for making good low-cost clear and pigmented sealers. Acrylics can also be used.

Freshly plastered surfaces, or new masonry, are high in moisture content and highly alkaline. An alkyd primer–sealer on these surfaces will quickly disintegrate because of saponification by the alkali. Such surfaces must cure 60 to 90 days before they can be safely sealed with an alkyd. Latex sealers are not affected by the alkali or water, and can be applied to these surfaces after two to three weeks of aging. However, the latex sealer may not be as efficient as an alkyd under the following conditions:

Overpowdery surface.
Overrotten plaster that has dried too fast or under high humidity conditions.
Efflorescence of soluble salts.

Formulation for a clear sealer is shown in Table 13.2. This is an acrylic latex with the addition of thickeners, coalescing agents, defoamers, and preservatives. It is a good sealer to apply over new plaster surfaces before applying alkyd wall finishes.

Formulation for a pigmented wall primer and sealer is shown in Table 13.3.

Primer–sealers may be tinted with the same color as the topcoat to improve hiding of the topcoat. Pigmented primer–sealers vary from 20 to 50% pigment volume, but for the best holdout of the topcoat should be below the critical pigment volume content (CPVC).

In many cases the latex wall finish is used as its own primer–sealer.

LATEX WALL PAINTS

The principal latexes used in wall finishes are polyvinyl acetates and polyacrylics. There are a wide range of raw materials that can be used, depending on the cost allowed. When the raw-material cost of the product is relatively high, the amount of hiding, as determined by the titanium dioxide levels and the volume solids of the paint, and the film properties that can be achieved

TABLE 13.2. CLEAR ACRYLIC LATEX SEALER.

	Pounds	Gallons
Acrylic latex[a] (44.5% solids)	409.0	46.20
Ethylene glycol	15.0	1.62
Water	148.0	17.76
Hexylene glycol	30.0	3.93
Ammonia hydroxide (28%)	1.2	0.16
2.5% Hydroxyethyl cellulose[b]	250.0	29.27
Defoamer[c]	2.0	0.26
Preservative[d]	0.8	0.10
	856.0	100.00
Weight/gallon 8.56 lb		
Total solids 22.0%		
Viscosity 82 K.U.		

[a]Rohm & Haas—AC-22.
[b]Union Carbide—OP-4400.
[c]Colloids Inc.—Colloid 600.
[d]Troy Chemical Corp.—Troysan 174.

will be at a higher level than when lower-cost raw-material costs are specified.

For flat wall paints the choice is between polyvinyl acetate latices or acrylic latices. Polyvinyl acetates have good color retention. Acrylic latices are more costly, but have excellent color retention and durability and achieve a higher degree of water resistance. Under normal usage as flat wall vehicles these latices have good alkali resistance and are permeable to water vapor.

The pigments normally used in latex paints are nonreactive and have essentially no solubility in water. The water-soluble-salt content of the pigments should be kept very low in order to minimize the introduction of reactive ions. Rutile titanium dioxide is the white pigment used for hiding, and there are specific grades used for interior flat wall paints which have special surface treatments. The titanium dioxide pigment is chosen to give maximum hiding, color, and stability.

Flat wall paints are low-gloss products, the low luster being achieved by the choice of appropriate extenders. These extenders include clays, calcium carbonates, silicates, diatomaceous earths, silicas, barytes, talcs, and so on.

Flat wall paints should have good application properties, good lapping and touch-up properties, good hiding, minimum burnishing, good stain removal, and good washability.

Ceiling paints are usually of higher PVC, since they are not required to have washability, but should have easy application with good uniformity of appearance, one-coat coverage, and good dirt resistance.

Good flat wall finishes should have one-coat coverage over most colors.

A formula for a low-cost flat wall finish is shown in Table 13.4. This is a polyvinyl acetate paint at 65% PVC using 1.35 pounds of titanium dioxide per gallon. This formula is also suitable for use as a ceiling paint.

In Table 13.5 is shown a formula for a high-quality wall finish—polyvinyl acetate/ethylene copolymer at 50% PVC containing 2.0 pounds titanium dioxide per gallon.

INTERIOR SEMIGLOSS AND GLOSS LATEX PAINTS

As mentioned previously, in film formation from a latex the film is formed from discrete polymer particles. Because of this it is difficult to attain glosses equivalent to those obtained from solvent systems where the binder is in solution.

TABLE 13.3. PIGMENTED POLYVINYL ACETATE PRIMER–SEALER

		Pounds	Gallons
Water		40.0	4.8
Dispersing agent[a] (10% solution)		10.0	1.20
Dispersing and wetting agent[b]		2.0	0.24
Polypropylene glycol[c]		2.0	0.23
Diethylene glycol monoethyl ether[d]		25.0	3.03
Dibutyl phthalate		18.0	2.06
Clay[e]		75.0	3.49
Titanium dioxide[f]		100.00	3.00
Disperse in a Cowles dissolver then add:			
Water		130.0	15.61
Methyl cellulose[g] (2% solution)		260.0	31.21
Preservative[h] (20% solution)		5.0	0.60
Polyvinyl acetate emulsion[i] (55% solids)		320.0	34.53
		987.0	100.00
Weight/gallon	9.87 lb		
Pigment volume concentration	24.4%		
Viscosity	79 K.U.		

[a]Rohm & Haas—Tamol 731.

[b]Antara Chemical Company—Emulphor EL 719.

[c]Dow Chemical—Polyglycol 1200.

[d]Union Carbide Corp.—Carbitol.

[e]Mineral & Chemical Corp.—ASP 400.

[f]E. I. du Pont de Nemours & Co.—Ti Pure R-901.

[g]Dow Chemical Company—Methocel 400C.

[h]Dow Chemical Company—Dowicide A.

[i]E. I. du Pont de Nemours & Co.—Elvacet 81-900.

TABLE 13.4. LOW-COST WALL FINISH: POLYVINYL ACETATE. (WHITE)

	Pounds	Gallons
Water	200.00	24.00
Preservative[a]	1.0	0.10
Hydroxyethyl cellulose[b]	5.5	0.65
Dispersant[c]	5.0	0.51
Potassium tetrapolyphosphate	1.5	0.10
Surfactant[d]	3.0	0.27
Ethylene glycol monoethyl ether[e]	10.0	1.26
Ethylene glycol	15.0	2.62
Defoamer[f]	3.0	0.27
Titanium dioxide[g]	135.0	4.29
Aluminum silicate[h]	175.0	9.54
Grind in Cowles and let down with:		
Polyvinyl acetate latex[i]	150.0	16.40
Water	330.0	39.99
	1037.0	100.00

Weight/gallon	10.4 lb
PVC	64.5%
Solids (wt.)	39.0%
Solids (vol.)	22.0%
Reflectance	0.930
Contrast ratio	0.985
Contrast ratio (toned)	1.000
Gloss (85°)	5.0

[a]Dow Chemical—Dowicil 100.
[b]Hercules Inc.—Natrosol 250 MR.
[c]Rohm & Haas—Tamol 731.
[d]GAF Corp.—Igepal CO-630.
[e]Eastman Kodak—Texanol.
[f]Diamond Shamrock—Nopco NDW.
[g]E. I. du Pont de Nemours & Co.—Ti Pure R-931.
[h]Burgess Pigment Co.—Optiwhite.
[i]Air Products & Chem. Co.—Flexbond 860.

There are several ways to improve the gloss of latex paints. The first is to use latices of smaller particle size, as these latices have improved coalescence because of greater interfacial compressive forces, and upon film formation the surface irregularities are less. However, smaller-particle-size latices sometimes cause stability problems. Also smaller-particle-size pigments may be used so that the latex particles have a better chance of surrounding the pigment particles; and a smaller amount of pigment is used, in order to lower the PVC. At lower PVC surface tack becomes a problem. If the hardness of the polymer is increased to overcome this problem, then the latex will not coalesce at room temperatures, and more coalescing agent must be used. Formulating a system of this type requires a series of compromises. Because of these problems most of the present commercial gloss latex paints are of the "semigloss" type.

TABLE 13.5. INTERIOR FLAT WALL FINISH: POLYVINYL ACETATE/ETHYLENE COPOLYMER AT 50% PVC. (WHITE)

	Pounds	Gallons
Water	110.0	13.22
Dispersing agent[a]	10.0	1.09
Potassium tripolyphosphate	1.0	0.05
Propylene glycol	25.0	2.90
Fungicide[b]	0.5	0.05
Defoamer[c]	1.0	0.13
Titanium dioxide[d]	200.0	6.02
Clay[e]	150.0	6.91
Methyl cellulose[f] (2%)	50.0	5.98
Whiting[g]	50.0	2.21
Diatomaceous silica[h]	25.0	1.30
Disperse in Cowles:		
2-amino 2-methyl-1-propanol[i]	4.0	0.51
Surfactant[j]	4.0	0.47
Methyl cellulose[f] (2%)	229.9	27.50
Water	24.0	2.88
Defoamer[c]	1.0	0.13
Polyvinyl acetate latex[k] (55)	254.8	28.64
	1140.1	100.00

PVC	50.1%	Freeze-thaw stability 16 hours at 0°/8 hr. @74°F	> 5 cycles
NVM by weight	51.0%	pH	9.0
by volume	32.8%		
Viscosity	83 K.U.	Scrub resistance D2486-T66	310 cycles
Contrast ratio	0.947	Theoretical coverage per gallon	527 sq ft

[a]Rohm & Haas—Tamol 731.
[b]Tenneco Co.—Nuodex PMA-18.
[c]Hercules Corp.—Defoamer 357.
[d]E. I. du Pont—Ti Pure R-901.
[e]Mineral & Chemical Co.—ASP-400.
[f]Dow Chemical Co.—Methocel 65GH-4000.
[g]Thompson Weinman & Co.—Snowflake.
[h]Johns-Manville—Celite 281.
[i]Commercial Solvents.
[j]Arolac Chemical Corp.—Abex 16S.
[k]E. I. du Pont—Elvace PB-3-1942.

An important application property of semigloss latex paints affecting their formulation is the need for adequate wet-edge time. In order to lap over previously painted surfaces without leaving brush or roller marks, the paints must stay wet longer than latex flat wall paints so that there is adequate reflow at the lapped areas. This is accomplished by using relatively high levels of propylene glycol in the formula. However, excessive amounts will cause sagging under certain conditions of temperature and humidity.

In general, semigloss latex paints have better color retention, retain their flexibility better, and are more alkali-resistant than solvent alkyd wall finishes. They have poorer hiding than alkyds and are more difficult to apply, since the application is more affected by temperature and humidity. They can be used in bathrooms and kitchens because of their good gloss uniformity and adhesion. New surfaces should be primed before application of a gloss or semigloss paint.

In Table 13.6 is shown a formula for an acrylic semigloss enamel meeting Federal Specifications TTP 1511A (GSA-FSS).

Stain resistance of wall finishes is evaluated by applying:

Mineral oil (9 parts) and Lampblack (1 part)
Lipstick
Crayon
Black china marker
Pencil
Ink

The stains should be removed by wiping with a nonabrasive household detergent without burnishing.

EXTERIOR HOUSE PAINT

Exterior paints may be applied to surfaces such as wood, masonry, and metal.

Wood, the most common surface, expands and contracts as its moisture content changes. (See Table 13.7.) With these changes in dimension a good paint must be flexible enough to stretch as the wood swells, even at low temperatures occurring during the winter season.

Masonry surfaces are dimensionally stable, but may be alkaline, especially when fresh. Masonry surfaces also contain water-soluble salts that will permeate through the paint surface giving a chalky deposit known as efflorescence.

Metal surfaces on the exterior of a house, such as iron, galvanized iron, aluminum, and copper, are sometimes painted. Iron surfaces usually have a mill oxide scale or rust from exposure. Loose particles should be removed by wire brushing, and then these surfaces should be painted with a penetrating solvent primer before application of a latex topcoat.

Galvanized iron causes adhesion problems with solvent paints. However,

TABLE 13.6. SEMIGLOSS LATEX PAINT: INTERIOR, ACRYLIC 27% PVC.* (WHITE)

Materials		Pounds	Gallons
Propylene glycol		70.0	8.12
Dispersant[a]		11.0	1.20
Defoamer[b]		2.0	0.27
Titanium dioxide—Rutile		250.0	7.30
Barytes		50.0	1.36
Disperse in Cowles and then add the following in thindown:			
Propylene glycol		100.0	11.60
Acrylic latex[c] (46.5%)		492.7	55.68
Defoamer[b]		2.0	0.27
Butyl cellosolve	premix	13.7	1.73
Surfactant[d]		2.0	0.23
Water	premix	50.0	6.00
Preservative[e]		2.6	0.11
Fungicide[f] (45%)		0.5	0.06
Water and/or hydroxyl ethyl cellulose[g] (2.5%)		57.8	6.07
		1104.3	100.00

Gloss	45%	Pigment volume content	26.8%
Solids (weight)	47.8%	Viscosity	75–80 K.U.
(volume)	32.2%	pH	8.5–9.0

*Meets Federal Specifications TTP-1511A.
[a]Rohm & Haas—Tamol 731.
[b]Nopco Chemical Co.—Nopco NDW.
[c]Rohm & Haas—Rhoplex AC-490.
[d]Rohm & Haas—Triton GR-7.
[e]Dow Chemical Co.—Dowicil 75.
[f]Rohm & Haas—Skane M-8.
[g]Hercules Chemicals Inc.—Natrosol 250 MR.

TABLE 13.7. EXTERIOR WOOD SURFACES: DIMENSIONAL CHANGE 20–0% H_20.

	Percent		
Wood	Radial	Tangential	Volume
Cedar	1.6	3.3	4.5
Redwood	1.7	2.9	4.5
Pondersosa	2.6	4.2	6.4
Southern pine	3.7	5.2	8.1

latex paints usually perform well on galvanized metal. Aluminum and copper, which have surface oxide films, can be painted with latex paints to give good results.

The early masonry finishes of the 1950s were based on styrene-butadiene latexes and were at lower PVC than interior wall finishes. In the late 1950s latex house paints based on acrylics were introduced for wood and masonry surfaces. These early paints were low in brushing viscosity so that too thin a paint film was applied. They also had poor adhesion over chalky surfaces.

About 10 to 20% of the pigment in latex house paints is zinc oxide for mildew resistance. Thickeners have been selected to increase the low shear viscosity so that an adequate film is applied (1 mil).

High-quality latex house paints generally contain 2.00 and to 2.50 pounds of titanium dioxide per gallon. This is a hiding pigment PVC of about 24%.

The polymer used as a binder for exterior paints should have good mechanical properties over a wide range of temperatures. An exterior paint applied over wood must be able to withstand the dimensional changes of the wood at low temperature without film rupture. The binder must exhibit some elongation and cold flow under strain. At higher temperatures encountered in the summer months, the binder must not display excessive flow or low tensile strength at very low strain rates. Coatings containing binders that exhibit excessive flow will tend to retain surface dirt even though the surface is not tacky to touch. The binder should not absorb excessive amounts of water, as this will soften and weaken the coating as well as causing cracking during freezing temperatures. An exterior house paint formula is given in Table 13.8.

Efflorescence present on either old or new masonry should be removed by washing with a 5% solution of hydrochloric acid and rinsing thoroughly with clean water. Mildew should be removed by scrubbing with a solution of sodium hydrochlorite (bleach) and trisodium phosphate.

It is risky to paint over old cement paints, lime, or whitewash coatings because these paints are powdery and are not a good base for later paints. It is best to remove these old coatings before repainting.

An exposure farm for evaluating exterior paint durability is shown in Figure 13.1.

FLOOR PAINTS

Floor paints are used for both interior and exterior applications on wood and concrete surfaces. The paint should be fast-drying and be ready for traffic within 24 hours after application. The floor paint should have good adhesion and abrasion resistance under wet conditions and have blister resistance when exposed to soap and water. Waterborne floor paints are not recommended for garage floors, as warm tires will soften the paint and cause lifting of the paint when the car is moved.

Surface preparation before floor paints are applied is very important. To obtain the best durability on interior concrete floors, they should be clean, relatively dry, and well cured before application of waterborne floor paints. The same conditions apply to exterior concrete surfaces such as patios. The

TABLE 13.8. EXTERIOR HOUSE PAINT: ACRYLIC MODIFIED WITH ALKYD (WHITE) (13%).

			Pounds	Gallons
Hydroxyethyl cellulose[a] (2½%)			85.0	10.37
Water			62.5	7.50
Dispersant[b] (30%)			10.5	1.05
Dispersant[c]			2.5	0.28
Potassium tripolyphosphate			1.5	0.07
Defoamer[d]			1.0	0.13
Ethylene glycol			25.0	2.69
Titanium Dioxide[e]			237.5	7.22
Zinc oxide[f]			50.0	1.07
Talc[g]			187.7	8.11
Grind the above materials in a Cowles and then add the following:				
Acrylic latex[h] (50%)			390.8	44.39
Long oil alkyd[i]			30.8	3.69
0.5% of 6% cobalt, 0.5% of 6% manganese, and 1.4% of 24% lead in alkyd				
Defoamer[d]			1.0	0.13
Tributyl phosphate			9.3	1.15
Propylene glycol			34.0	3.95
Fungicide[j] (45%)			2.0	0.23
Ammonium hydroxide (28%)			1.0	0.13
Water			65.3	7.84
			1197.4	100.0
Pigment volume content	40%	Viscosity	72–76 K.U.	
Solids: volume	41%	pH	9.5	
weight	48.5%			

[a]Hercules Chemical—Natrosol 250 MR.
[b]Rohm & Haas—Tamol 850.
[c]Rohm & Haas—Triton CF-10.
[d]Nopco Chemical—Nopco NZX.
[e]E. I. du Pont—Ti-Pure R-960.
[f]American Zinc Sales Co.—AZO-11.
[g]International Talc Co.—Abestine 3X.
[h]Rohm & Haas—Rhoplex AC 388.
[i]Ashland Chemical Company—Aroplaz 1271.
[j]Rohm & Haas—Skane M-8.

FIGURE 13.1. Exposure farm for evaluating exterior durability of paints.

adhesion of coatings has been improved by pretreating the surface with a 10% solution of phosphoric acid and then rinsing with water. Waterborne paints will not adhere over greasy or oily surfaces. If they are applied over old paints, the gloss must be removed by sanding.

A formulation for an acrylic-epoxy floor enamel is given in Table 13.9.

STAINS FOR EXTERIOR WOODS

Latex-based stains are becoming increasingly popular as finishes for exterior woods. Exterior durability tests have shown that latex-based stains are superior to oil-based stains for all-around weather resistance. These stains are available as solid colors or semitransparent coatings.

Latex stains preserve the woods, show natural texture, resist blistering, chipping, and peeling, and are fade-resistant. They can be applied by pad applicator, roller, brush, or spray to siding, shingles, shakes, decks, doors, fences, furniture, and planters.

Solid-color stains, which account for most of the sales, are designed to create a fashionable, uniform appearance by hiding grain imperfections and color variations in new wood or badly weathered wood surfaces, while permitting the texture of the wood to remain visible. Semitransparent stains are used to protect and enhance the appearance of natural grain and texture of fine-quality woods.

The application viscosity, pigment volume concentration, and total solids are the important factors in formulating good latex stains.

Solid-color stains in the 60 to 75 K.U. viscosity range provide the best application, lap-in, and penetration. Pigment volume concentration is especially important in attaining grain crack resistance; also lap-in characteristics are impaired if the PVC is too low. Around 30% PVC seems to be the optimum for maintaining crack resistance and lapping properties. A total solids content of 40% is the maximum that can be used without adversely affecting penetration and grain crack resistance.

The solid-color stains resemble house paints in opacity. Deep tones and medium shades are based on oxide pigments. Semitransparent types may be factory-tinted or supplied as bases for use with universal colorants.

TABLE 13.9. FLOOR PAINT: ACRYLIC MODIFIED WITH EPOXY (GRAY).

	Pounds	Gallons
Dispersing agent[a] (25%)	7.5	0.82
Dispersing agent[b]	2.0	0.20
Defoamer[c]	2.0	0.20
Water	80.4	9.73
Titanium dioxide[d]	228.6	6.54
Lampblack dispersion	30.0	2.54
Grind in Cowles and then add the following in the letdown:		
Water	26.1	3.16
Propylene glycol	54.6	6.17
Preservative[e]	1.0	0.10
Acrylic latex[f] (46%)	485.4	54.76
Epoxy emulsion[g] (50%)	49.6	4.96
6% Cobalt drier	0.2	0.02
25% Lead drier	1.1	0.11
Aluminum oxide[h]	24.8	0.67
Butyl cellosolve	24.4	3.24
Hydroxyethyl cellulose[i] (3%)	67.8	8.19
	1085.5	102.4

Solids: weight	48.2%
Pigment volume content	21.7%
Viscosity	60–65 K.U.
Gloss (60°)	39

[a]Rohm & Haas—Tamol 731.
[b]Rohm & Haas—Triton CF-10.
[c]Colloids Inc.—Colloid 600.
[d]E. I. du Pont—Ti-Pure[R]-R-900.
[e]Tenneco Chemicals Inc.—Super Ad-It.
[f]Rohm & Haas—Rhoplex AC-61.
[g]Ciba Products Co.—Araldite DP-624.
[h]Exolon Co.—SD-No. 220 Mesh.
[i]Union Carbide Chemical Co.—WP-4400.

The addition of drying oil or oil-modified alkyd aids penetration and adhesion, especially over previously painted and poorly prepared surfaces. Modification in the 10–20% range based on vehicle solids gives the best results.

A typical formula for a solid-color stain is shown in Table 13.10.

Latex stains are intended for application over those woods commonly used for siding, including cedar, redwood, cypress, fir, pine and exterior grades of textured or rough-sawn plywood. Soft porous woods in general yield best

TABLE 13.10. LATEX SHINGLE STAIN: VINYL ACRYLIC. (RED)

	Pounds
Water	250.0
Hydroxyethyl cellulose[a]	3.0
Dispersant[b]	4.5
Surfactant[c]	3.0
Potassium tripolyphosphate (KTPP)	1.0
Antifoam[d]	1.0
Ethylene glycol	10.0
Preservative	2.0
Titanium dioxide—Rutile[e]	25.0
Aluminum silicate[f]	50.0
Zinc oxide	50.0
Silica[g]	25.0
Black oxide[h]	15.0
Red oxide[i]	65.0
Grind in Cowles and then add the following in the letdown:	
Butyl carbitol	15.0
Water	195.0
Vinyl acrylic latex[j] (55%)	305.0
Antifoam[d]	2.0
	1021.5

Solids: weight	40.0%	Weight per gallon	10.28 lb.
Pigment volume concentration	30.0%	Viscosity	61.0 K.U.

[a]Union Carbide Corp.—QP-52,000.
[b]Rohm & Haas—Tamol 850.
[c]GAF—CO-630 Surfactant.
[d]Witco Chemical Co.—Balab 748.
[e]E. I. du Pont—Ti-Pure R-960.
[f]Indusmun—Minex 4.
[g]Johns-Manville—Celite 281.
[h]Pfizer—BK 5099.
[i]Pfizer—RO 7097 Kroma.
[j]Union Carbide Corp.—Ucar 366.

results. Latex stains are self-priming and should be applied directly to new or previously stained and weathered surfaces. They should not be used over layers of old paint or any type of sealed, nonporous surface. Latex stains produce a durable finish and should never be topcoated with a clear varnish or sealer of any kind. In most cases, two coats of latex stain will give satisfactory service. Allow four to six hours of drying time between coats.

Coverage will range from 200 to 400 square feet per gallon, depending on the porosity and texture of the surface. Unlike oil stains, latex stains can be applied to damp surfaces. In fact, with very porous wood the surface can be hosed to remove dirt before the first coat of stain is applied. In hot, dry weather, prewetting slows the drying rate and helps prevent an uneven appearance in lap and touch-up areas. Since latex stains are at relatively low viscosity, they should be stirred before, and frequently during applications.

PROBLEMS WITH LATEX PAINTS

There are certain problems associated with the formulation, manufacture, storage, and application of latex paints. Some of these problems are:

I. Coagulation or Viscosity Increase in Storage

Cause	*Remedy*
(a) Insufficient or inefficient wetting and dispersing agents.	(a) Use types and amounts recommended by latex suppliers.
(b) Presence of reactive pigments.	(b) If zinc oxide or extended titanium dioxide is present, special surfactant systems must be used.
(c) pH drift.	(c) Buffer system for better pH control. Incorporate alkaline pigment or switch to thickener with wide latitude for pH difference.
(d) Excess strong solvent.	(d) Decrease amount of strong solvent.
(e) Latex with poor compounding stability.	(e) Change latex or check with the latex supplier.
(f) Prolonged exposure to elevated or freezing temperatures.	(f) Change storage conditions or reformulate.

II. Viscosity Decrease in Storage

Cause	*Remedy*
(a) Bacterial or enzyme degradation of thickeners.	(a) Improve plant housekeeping; change to more resistant thickeners.
(b) Improper wetting system.	(b) Increase the amount of wetting agent, or change to more efficient system.

(c) pH drift with pH-sensitive thickener.	(c) Buffer paint to maintain pH, or switch to thickeners unaffected by pH.

III. Freeze-Thaw Instability

Cause	*Remedy*
(a) Too high level of coalescing agent.	(a) Use the minimum amount of coalescing agent needed to obtain satisfactory performance.
(b) Insufficient glycol.	(b) & (c) If paint fails by pigment flocculation, an increase in surfactant will help. If failure is by coagulation, an increase in either surfactant, glycol, or protective colloid, or all three, may be required.

IV. Fisheyes or Crawling on Application

Cause	*Remedy*
(a) Improperly mixed defoamer. (b) Too much defoamer or wrong type for rest of system. (c) Insufficient wetting agent. (d) Incompatible hydrophobic constituent, improperly emulsified. (e) Paint applied over contaminated surfaces.	(a–e) Make appropriate changes.

V. Poor Color Uniformity over Sealed and Unsealed Surfaces

Cause	*Remedy*
(a) Insufficient coalescing agent.	(a) If color is lighter on unsealed surface than on sealed surface, usually more coalescing agent or a stronger one is required.
(b) Excessive amount of coalescing agent.	(b) If color is darker on the unsealed surface, less coalescing agent is needed.
(c) Wrong thickener for this particular latex.	(c) Change thickener.

VI. Poor Sheen Uniformity

Cause	*Remedy*
(a) Extreme variation in substrate texture.	(a) Proper surface preparation.

(b) Tiny air bubbles trapped in coating.	(b) Change defoamer or increase level.
(c) Incorrect extender system.	(c) Extenders that flatten without orienting under brushing or rolling should be used. Generally a combination of two or more extenders is best.
(d) Incorrect coalescing agent.	(d) See V, (a) and (b).
(e) Incorrect thickener.	(e) Change thickener.
(f) Poor film formation.	(f) Insufficient coalescing agent was used, or film may have dried at low temperatures or with too-rapid drying; make appropriate changes.

VII. Color Variation on Storage

Cause	*Remedy*
(a) Unsuitable surfactant or incorrect amount.	(a) Follow latex supplier's information.
(b) Pigment flocculation resulting from pH drift effect on thickener.	(b) Switch to a thickener not sensitive to pH changes.
(c) Incomplete color development.	(c) Add thickener.
(d) Test paints dried under varying conditions of temperature and air circulation.	(d) Dry test paints under standard temperature and conditions.

VIII. Poor Leveling

Cause	*Remedy*
(a) Excessive thixotropy with fast of viscosity.	(a) If thixotropic or nondrip finish is desired, leveling will be poor. If thixotropy is not desired, add more surfactant. The presence of high-water-demand pigments will give poor leveling.
(b) Water loss too rapid.	(b) Increase glycol content; reduce amount of high-water-demand pigments. Increase amount of thickener by using a lower-viscosity grade.
(c) High viscosity.	(c) Use less thickener.

IX. Poor Color Development When Tinting

Color	*Remedy*
(a) Insufficient surfactant or wrong type or system.	(a) Use correct surfactant.
(b) Presence of low gel grade of hydroxypropyl methylcellulose in paint tinted while warm.	(b) Change to higher gel grade hydroxypropyl methylcellulose.

X. Poor Touch-Up

Cause	*Remedy*
(a) Poor film formation.	(a) See VI (f).
(b) Poor color development.	(b) See VII, (c).
(c) Poor color uniformity.	(c) See VII.
(d) Poor sheen uniformity.	(d) See VI.

XI. Poor Wet Abrasion Resistance

Cause	*Remedy*
(a) Poor film formation.	(a) See VI, (f).
(b) Excessive surfactant.	(b) Reduce level of surfactant.
(c) Excess of high-water-demand pigments.	(c) Correct amount of pigments.
(d) Poor adhesion to substrate.	(d) Prepare surface properly.
(e) Incorrect use of coalescing or plasticizers.	(e) Follow latex supplier's directions.

XII. Poor Hiding for Titanium Dioxide Level

Cause	*Remedy*
(a) Flocculation of titanium dioxide.	(a) See III, (a), (b), and (c).
(b) Improper manufacturing procedure.	(b) Follow latex supplier's recommendations.
(c) Reaction of ingredients with anionic dispersants.	(c) Find offending ingredient.

XIII. Loss of Hiding on Aging

Cause	*Remedy*
(a) Flocculation of titanium dioxide.	(a) See III, (a), (b) and (c).
(b) pH drift in pressure of sensitive thickeners.	(b) See VI, (b).
(c) Unsuitable surfactant or dispersant or both.	(c) Make changes.
(d) Frozen paint.	(d) See III.

XIV. Poor Enamel Holdout

Cause	*Remedy*
(a) Poor film formation.	(a) See VI (f).
(b) Presence of high-vehicle-demand pigments in paint applied over porous surfaces.	(b) Prepare surface properly.

XV. Pinholing in Dried Film

Cause	*Remedy*
(a) Excess foam in paint.	(a) See IV, (a) and (b).
(b) Chemical reaction in storage, causing liberation of gas.	(b) See I, (b) and II, (a).
(c) Paint applied over extremely porous surfaces.	(c) Prepare surface properly.

XVI. Peeling, Poor Adhesion

Cause	*Remedy*
(a) Paint applied over chalky surfaces.	(a) Proper surface preparation—use penetrating primer.
(b) Paint applied over old paint having poor adhesion.	(b) Scrape old paint, and prepare surface properly.

14
Maintenance Paints

Maintenance paints are high-performance coatings that are applied to bridges, storage tanks, silos, steel structures, piping, and other equipment. These coatings must protect the surface from attack by industrial atmospheres which may contain chemical fumes and have a high moisture content. The paint surfaces should also be resistant to spillages and abrasion. Ability to provide decoration is important but is secondary to that of protection.

Maintenance coatings are expected to give many years of service under severe conditions. They are relatively thick coatings as compared to other types of coatings.

TYPES OF PROTECTION

There are two mechanisms of paint protection: barrier and galvanic (sacrificial).

Conventional paints form a resistant barrier between the environment and the material being protected. Satisfactory protection requires that the coating be applied and maintained as a continuous film without skips, pinholes, or other breaks in the coating that would expose the substrate to the environment.

The other mechanism, that of sacrifical protection, is well known and has been very effective, for example, on the use of galvanized coating to protect steel. Zinc-rich paints similarly provide coatings for steel which are substantially metallic zinc, the dried film containing 85–95% zinc metal by weight. When a metal such as zinc is in electrical contact with a metal more noble in the electromotive series such as iron, a galvanic cell is established, and the electric current will flow when an electrolyte contacts the metallic couple. The zinc becomes the anode and the iron the cathode. In a zinc-rich paint–steel substrate system, the electrolyte is provided by atmospherically contaminated water, rain, or dew. Under these conditions, electric current will cathodically protect the substrate from corrosion, the zinc being preferentially corroded and wasted away in the process. It is for this reason that galvanic protection is sometimes called sacrificial protection. Thus in galvanic protection it is not essential as with barrier protection to have pinhole-free coatings, as protection will be provided at any exposed areas by electrical action.

FAILURE MECHANISM

The most common substrate in industrial maintenance applications is steel. A coating over steel can fail in many ways. If there are any breaks or discontinuities in the barrier paint, film protection will be lost at those points, and the environment can enter and attack the substrate. The result of such an attack by humidity, moisture, or most chemicals in the presence of moisture is rusting. On the average one unit volume of steel, in the process of rusting, will yield 20 volumes of rust. This expansion will lift the edges of the coating and finally cause total paint failure.

COATING THICKNESS AND NUMBER OF COATS

Since the chances of obtaining continuity increase with increasing thickness of the coating, there is a minimum thickness below which it is difficult to obtain protective continuity because of surface irregularities. For most applications 5 mils (dry) seems to be the minimum film thickness that should be used.

While it is possible with some paints to apply a 5-mil coating in one application, it is preferred to apply three thinner coats to minimize holidays, skips, pinholing, and so on. A well-known rule of maintenance painting is "A minimum of 5 mils with a minimum of three coats."

PREVENTIVE MAINTENANCE

Preventive maintenance entails periodic inspection of the coatings. It is necessary that the protective barrier be maintained throughout the entire service life of the coating. Reduction of thickness by chalking, weathering, or service conditions of mechanical injury or abrasion may reduce the barrier thickness below the required 5 mils. Thicknesses over metals can be checked easily with a dry film thickness gauge, which does not harm the film.

SURFACE PREPARATION

In all cases it is necessary to remove loose rust and other incompatible residues. In mild exposures and with proper selection of primer, adequate performance usually can be obtained by painting over adherent rust. For severe exposure of prying corrosion types, a surface preparation of sand blasting is mandatory. As a general rule painting over mill scale will lead to premature failure caused by moisture permeating the coating and oxidizing the mill scale to more voluminous rust. Surfaces for application of waterborne paints must be cleaner than for solvent finishes, as waterborne paints will not wet oily surfaces properly.

FIGURE 14.1. Maintenance painting with emulsion paints.

PRIMERS

A primer is a coat that adheres to the substrate and forms a good base for the subsequent coats to adhere to.

Inhibitive pigments are often incorporated into primers to retard steel corrosion when it is exposed to humid environments. These inhibitors are not effective in alkaline exposure and may actually hasten the breakdown of the coating. The thickness of the primer need only be enough to accomplish the purpose of providing mutual primer adhesion to the substrate and the topcoats; usually 1.0 to 1.5 mils is adequate.

INTERMEDIATE COATS AND TOPCOATS

In maintenance painting, intermediates are usually the same composition as the finish coats, although preferably they differ in color or tint in order to provide contrast between coats. Since the primer is applied at about 1 mil dry film thickness, to obtain a total thickness of 5 mils two an intermediate coat and a top coat of about 2 mils are applied. In order to obtain this thickness, the intermediate-coat formulation sometimes is "bulked" with fillers, increasing its permeability. In such cases it is necessary to have an adequate thickness of topcoat applied to seal such intermediate coats.

WATERBORNE MAINTENANCE PAINTS

In 1963 the first acrylic emulsion paints for bare steel were introduced. General acceptance was slow, since many maintenance engineers believed that "Water rusts steel so latex paints will never work."

However, many of the raw-material suppliers have used waterbonre maintenance paints on their own facilities with good results, with service records up to 10–12 years. The coatings were applied to steel structures, support beams, steel tanks, and water towers in many locations under a variety of weather conditions. Aluminum storage silos were painted with a system consisting of a vinyl butyral wash primer and followed with an acrylic latex topcoat, with very good results.

FLASH RUSTING

One of the problems a painter may encounter in applying waterborne coatings to a metal surface is flash rusting. This usually occurs under conditions of high humidity and slow drying. Dissolved oxygen in the paint may be one of the causes.

There are several ways to control flash rusting. One is by the use of rust-inhibitive pigments at a low level of 0.5 to 1.0 pound per gallon of paint; molybdenum-containing pigments, lead chromate, and zinc phosphate are among the effective ones. Rust inhibitors in the range of 0.25% such as ammonium benzoate, sodium benzoate, sodium nitrate, potassium nitrate, sodium phosphate, amine methyl propanol (AMP), and morpholine are effective. The salts work via the anodic mechanism or by buffering the paint on the basic side. The amines maintain basic conditions during drying. A test for flash rusting is to coat a properly cleaned steel panel and dry it at a temperature of 80°F and 90% relative humidity. If flash rusting results, light-colored coatings will become dark or will actually exhibit rust spots. If the coating is dark, it must be stripped off with a solvent and the panel examined for rusting. Flash rusting is more prevalent with polyvinyl acetates than with acrylics.

Early rusting is another type of rusting, which shows up as rust discoloration of a latex maintenance paint and may occur within several days after the coating has dried. Early rusting is favored by thin paint films (less than 1.5 dry mil) when painting is done at temperatures near 50°F and under conditions of high humidity.

It takes several days to complete film formation in a latex coating. During the drying of latex coatings individuals latex particles contact each other, adhere, and then are forced together by surface tension. A third stage occurs later and continues film formation for several days. There is a gradual diffusion of polymer chains between latex particles to complete the film formation sequences. Early rusting, when it occurs, takes place after the initial film formation but before final film consolidation is completed.

A good latex maintenance paint will form good films at relatively high humidities if the ambient temperature is 70°F or higher. The latex used should be fine-particle size (i.e., as low as 0.15 micron).

CHEMICAL AND SOLVENT RESISTANCE

Acrylic maintenance paints show very good resistance to acids, but, as might be expected, strong solvents such as aromatic hydrocarbons, ketones, and esters soften them; and, as is typical of polymers containing ester linkages, they have only fair alkali resistance. Solvent resistance can be improved by using latexes that cross-link upon drying.

WATERBORNE SYSTEMS

Several waterborne paint systems have been used successfully for maintenance painting of steel. They are:

1. Inorganic zinc
2. Inorganic zinc/latex topcoat
3. Inorganic zinc/water-based topcoat
4. Latex metal primer/latex intermediate/latex topcoat
5. Asphalt emulsions

Latex maintenance paints are unique. They function entirely differently from solvent maintenance paints, which offer corrosion resistance primarily by acting as a moisture or vapor barrier. Latex coatings, on the other hand, while having poorer moisture vapor transmission characteristics, obtain much of their corrosion-inhibiting properties through an anodic complexing mechanism of inhibiting pigment. Maintenance latex paints are not simply trade sales paints with a corrosive inhibitor, but are formulated with latexes specifically designed for use on metal.

For the anodic mechanism to function properly, the corrosion-inhibitive pigment must be very slightly soluble in water. The anion portion of the inhibitor may then complex with soluble iron corrosion products to give an insoluble passivating film. For example, Busan* 11 M-1, barium metaborate ($BaB_2O_3H_2O$), is 0.3 to 0.4% soluble in water, and presumably the metaborate anion, B_2O_4, can complex with iron or other cations. The need for both slight water solubility of the corrosion-inhibiting pigment and the lack of water sensitivity of the paint film are contradictory properties, so that the balance chosen is very important.

In addition to the potential problem with moisture sensitivity, soluble inhibitive pigments present the possible problem of shocking the latex system. A major portion of latex maintenance paint formulating revolves around the following parameters; the need for a good corrosion inhibitor with minimum water sensitivity, selecting a dispersing system for adequate stability, and selecting the proper latex.

*Buckman Laboratories.

Corrosion-Inhibiting Pigments

As mentioned previously, the selection of corrosion-inhibiting pigments for use in latex coatings is somewhat different from that for solvent systems, as performance in latex coatings depends on the pigments' partial water solubility.

In one study of corrosion-inhibiting pigments in latex coatings on steel, lead silicochromate, strontium chromate, zinc phosphate, calcium borosilicate, and calcium zinc molybdate were evaluated. They were tested in acrylic, polyvinyl acetate, and vinylidine chloride latices at PVC's of 35%. Titanium dioxide was also included in the series as a noninhibitive pigment. Panels were subjected to salt spray and humidity tests. Results are shown in Table 14.1.

Other Pigments and Extenders

Zinc oxide appears to improve the performance of most waterborne maintenance paints. A finely divided French process grade zinc oxide is preferred.

Calcium carbonate is used as precipitated grades low in water-solubles. As corrosion occurs more readily at low pH, the slightly basic calcium carbonate helps in corrosion prevention.

Silica may replace calcium carbonate. In high-humidity tests, systems with silica tend to blister less over clean steel. However, greater rust bleed has been noted with silica than with calcium carbonate in salt spray tests.

Mica is used at relatively low levels with the platelet-type particles reducing moisture transmission through the film

Application

Most waterborne paints are readily applied by brushing or rolling; however, this is not always practical for maintenance painting. Most waterborne mainte-

TABLE 14.1. CORROSION-INHIBITING PIGMENTS IN LATEX COATINGS.[a]

Pigment	Salt spray	Humidity	Overall
Lead silicochromate	6	4	10
Strontium chromate	4	6	10
Zinc phosphate	3	5	8
Calcium borosilicate	6	2	8
Calcium zinc molybdate	3	3	6
Titanium dioxide	1	1	2

[a]6 = most inhibitive; 1 = least inhibitive.

nance coatings are applied by spraying, either air spray or airless spray. The waterborne paint will normally be applied at somewhat lower fluid and atomization pressures than for solvent coatings.

Airless spray is fast becoming the most popular method of applying industrial maintenance paints. Latex paints can be applied at higher solids with high film build and less sagging than with solvent paints. Overspray, misting, and fogging are reduced by more than 50% as compared to air spraying.

Inhibitive Primers

Table 14.2 gives a formula for a red oxide inhibitive primer for low-cost shop coat and general maintenance use.

TABLE 14.2. LATEX MAINTENANCE INHIBITIVE PRIMER (STYRENE-ACRYLIC). (RED)

		Pounds	Gallons
Water		240.0	28.7
Dispersant[a]		8.0	0.8
Wetting Agent[b]		5.0	0.6
Ethylene glycol		20.0	2.2
Defoamer[c]		5.0	0.6
Thickener—hydroxyethyl cellulose[d]		3.0	0.3
Mica (325 mesh WG)		35.0	1.5
Inhibitive pigment—barium metaborate[e]		50.0	1.8
Red Oxide[f]		150.0	5.0
Calcium carbonate		75.0	3.3
Disperse in Cowles then add:			
Styrene acrylic latex[g] (46%)		420.0	49.6
Alkyd[h] (100%)		45.0	5.3
Cobalt drier (6%)	premix	1.0	0.16
Zirconium drier (6%)	premix	1.0	0.14
		1058.0	100.0
Solids (wt.)	53.0%		
Solids (vol.)	38.7%		
Viscosity	68 K.U.		
PVC	30%		
Weight/gallon	10.6 lb		

[a]Rohm & Haas—Tamol 850.
[b]GAF—Igepal CA-630.
[c]Witco Chemical—Bubble Breaker 745.
[d]Union Carbide Corp.—Cellosize QP 4400.
[e]Buckman Laboratories—Busan 11M.1.
[f]Pfizer, Inc.—N2060F.
[g]Union Carbide—Ucar 4341.
[h]McCloskey Varnish Co.—Varkyd 515-100.

Maintenance Gloss Topcoat

Table 14.3 gives a formula for an acrylic maintenance gloss topcoat.
A formula for a zinc-rich coating appears in Chapter 16.

TABLE 14.3. MAINTENANCE GLOSS TOPCOAT: ACRYLIC 15% PVC. (WHITE)

Material		Pounds	Gallons
Water		83.0	10.0
Potassium tripolyphosphate		2.0	0.20
Wetting agent[a]		2.0	0.23
Defoamer[b]		3.0	0.39
Preservative[c]		1.5	0.20
Titanium dioxide[d]		210.0	6.10
Ethylene glycol		20.0	2.15
Disperse in Cowles and add the following:			
Acrylic latex[e] (45%)		660.0	76.00
Butyl carbitol		16.0	2.00
Ammonium hydroxide (28%)		2.0	.30
Thickener[f] (11%)		10.0	1.19
Water		12.0	1.44
		1021.5	100.2
Solids (wt.)	51.0%		
Solids (vol.)	39.3%		
PVC	15.5%		
Viscosity	68 K.U.		
pH	8.3		
Gloss 60°	70		

[a]GAF—Igepal CA-630.
[b]Witco—Bubble Breaker 748.
[c]Merck—Merbac 35.
[d]NL Industries—Titanox 2020.
[e]Union Carbide Corp.—Ucar 4358.
[f]Rohm & Haas—Acrysol G110.

15
Industrial Coatings

The adoption of waterborne coatings for industrial applications has been slower than in the trade sales area. Industrial-type finishes are defined as coatings applied on the production line to various products such as automobiles, appliances, furniture, coil coatings, building materials, drums, and so on. Industrial finishes are sometimes known as chemical coatings and are formulated specifically for each application. Surfaces coated may be metal, wood, plastic, rubber, fabric, and paper. In most cases these are baking-type finishes.

Waterborne industrial paints have been used for the past 20 years or more for such industrial applications as acoustical tile, steel strapping, hardboard primers, and so forth.

With the onset of government regulations to reduce the amount of organic solvents released to the atmosphere, there is forced incentive to change from solvent paints to some other coating systems. In many cases waterborne systems are the most economical way to go. (See Table 15.1.)

In most cases the costs of waterborne coatings are higher than the costs of comparable solvent coatings. (See Table 15.2.)

Table 15.3 gives comparative data for solvent emission of various coating types.

CONVERTING TO WATERBORNE COATINGS

In spray, dip, or flowcoat operations the conversion to waterborne coatings will likely be the first option considered for many facilities because of the possibility that these coatings can be applied essentially with existing equipment.

Converting to waterborne coatings provides a potential decrease in toxicity and flammability.

Waterborne coatings may be thinned with water, and coating equipment cleaned or flushed with water. When they dry, however, waterborne coatings must be cleaned with organic solvent, as they are no longer soluble in water. The addition of ammonia to the cleanup mixture sometimes helps.

While the heat of vaporation of water is greater than that of organic

TABLE 15.1. WATERBORNE COATINGS VS. SOLVENT COATINGS.

	Emulsion paints	Water-soluble paints	Solvent paints	High solids 70%
After burners	Generally not required	Generally not required	Generally required	Generally not required
Lower explosive limit in oven	Greatly reduced	Much reduced	—	Very slightly reduced
Air pollution	Meets most if not all regulations	Meets most regulations	—	Will not meet most regulations
Insurance	Generally reduced	Generally reduced	—	No change
Energy requirements for cure	Up to 40% lower	Up to 25% lower	—	No change
Chemical and physical properties	Equal	Equal	—	Equal
Toxicity considerations	Generally much better	Generally better	—	No different or more severe
Use of costly solvent	Greatly reduced	Much reduced	—	Much increased
Cost/ft^2/mil	Slightly higher	Slightly higher	—	Higher
Application electrostatic spray	Very difficult	Difficult	—	Difficult
Dip or flow coat	Very difficult	Difficult	—	Usually not practical

TABLE 15.2. RELATIVE PAINT COST PER GALLON.

Coating Type	Conventional solvent paint	Waterborne paint
Alkyd	1.00	1.20–1.25
Polyester	1.00	1.25–1.30
Acrylic	1.00	1.10–1.15

solvents, the amount of air flow in the oven may be reduced to stay below the explosive limit.

There are several disadvantages of waterborne paints as compared to conventional solvent paints, which require that the production line be modified in many ways.

In spray application closer attention to temperature, humidity, gun-to-surface distances, flash-off time, and so on, should be observed, as compared to conventional solvent paints. In electrostatic spray the entire system must be isolated.

Waterborne coating applied by conventional dip and flow coating application need to be monitored more closely because of their sensitivity to pH changes, solvent loss, and so on.

PRODUCTION LINE CHANGES

1. *Lengthening of Flash Tunnels and Ovens*. Waterborne coatings require a longer flash tunnel prior to curing. Temperatures must be raised more slowly in order to evaporate the water slowly enough to avoid pitting of the coating. This necessitates longer ovens, which in turn may force equipment relocation.

2. *Cleanliness*. Waterborne coatings do not "dry to touch" as quickly as

TABLE 15.3. TYPICAL SOLVENT EMISSIONS OF COATING PER 1000 SQ/FT @1.0 MILS.

Type coating	NVM	% Solvent	Pounds solvent
Emulsion coating	55	< 5	0.50
Water-soluble coating (20% cosolvent)	45	11	1.46
Conventional solvent coating	45	55	9.84
High solids coating	70	30	4.00
High solids coating	86	14	1.61

conventional solvent paints. Thus they are more susceptible to dirt pickup. This necessitates filtration of incoming spray-booth air. Overhead conveyors may be unacceptable because of the potential for dropping dirt on newly painted parts.

3. *Humidity and Temperature Conditioning.* Proper temperature and humidity conditioning of the makeup spray-booth air is vital. If the humidity is too high or the temperature too low, the coating will sag on vertical surfaces. If the humidity is too low and the temperature too high, the coating will not flow, causing "orange peel." While spray primer applications may not be as critical, topcoats applications need careful control. Water can be removed from air by chemical or mechanical means. The chemical means involves the use of a hydroscopic solution; the mechanical means involves the use of a refrigeration cycle.

4. *Corrosion Control.* The pipes used to convey solvent paints from the central mixing areas to the application points are not suitable for waterborne paints and should be replaced with corrosion-resistant materials. In many cases stainless steel or polyvinyl chloride is used. The lifetime of carbon steel spray booths may also be lessened when waterborne coatings are used. Other components of the painting system may have to be changed.

5. *Adequate Rinsing.* In chemical metal treatments the metal must be rinsed well, as any chemicals carried over to dip or flow coating operations may cause instability in the waterborne paint.

6. *Cleanup.* Compared to overspray from solvent coating, waterborne-paint overspray is finer and does not dry in air before being drawn through the particulate collector, thus causing increased cleanup problems and costs. It is more difficult to clean up spray booths, conveyors, and so forth.

7. *Pump Maintenance.* Because of an apparent reduction in lubricity, pumps and other equipment with mechanical seals require more frequent repair.

8. *Sludge Handling.* Waterborne paints do not harden in water-wash particulate collectors; so sludge handling is more difficult.

9. *Line Shutdown.* Because of the potential for rusting and dirt pickup, articles coated with waterborne coatings cannot be left wet overnight or even during shift changes. The line must have facilities for carrying painted parts through the oven after the line shuts down.

The cost of converting to waterborne paints for an existing plant will vary. A major variable will be the age of the existing coating equipment. If it is very old, it may be better to build entirely new spray booths and ovens.

If the coating equipment is still relatively modern, however, converting will entail lengthening of the ovens and modification of spray booths and conveyors. For conversion, capital costs are about one-half those of building a new plant.

INDUSTRIAL FINISH PRODUCTS

The following products may be finished with waterborne coatings:

Automotive products	Exterior body	Primer surfacer Enamel topcoat
	Underside parts	Frame, springs, brackets, engine block, wheel rims, misc. parts
Farm and road equipment	Metal surfaces	Primer sealer Enamel Topcoat
Appliances	Air conditioners Refrigerators, freezers Washers and dryers	Primer Topcoat
Coil coating formed after painting	Siding Venetian Blinds Shelving Roof decking, etc.	Primer—in some cases Topcoat
Can coatings	Linings Base coats Side seam spray End seal compound	Interior coating Exterior Exterior Exterior
Fabric coatings		
Acoustical tile	Fiberglass Cellulosics Gypsum	Topcoat
Hardboard	Cellulosics Asbestos cement	Primer Topcoat
Wood	Paneling Furniture	Stains Sealer, filler Topcoat
Misc. industrial products	Steel strapping Drums	Topcoat Topcoat and liner

PAINTING PLASTICS

Plastics are painted for decorative and functional purposes. For example, a purely decorative application is a simulated woodgrain finish on plastic. A functional coating may be a highly conductive coating on the inside of a cabinet for electrical shielding. In many cases a piece of equipment may be made up of metal and plastic exterior parts, and the paint gives the overall assembly a uniform appearance.

The following is a list of the most paintable plastics:

Acrylics	Polyethylene
Cellulosics	Polypropylene
Epoxy	Polystyrene
Nylon	Polysulfone
Phenolics	Polyurethane
Phenylene oxide type	Polyvinyl chloride
Polycarbonate	Styrene acrylonitrile
Polyester	

If the plastic contains a plasticizer, it may migrate to the surface and soften the coating; the mold release on the surface of the plastic may cause trouble. Many products on the market are termed "paintable mold releases"; they may be used successfully with some types of paints but not with others. Zinc stearate is probably the worst of the mold releases. It is incompatible with paint and very insoluble and difficult to remove. Generally it is removed by scrubbing with a brush, soap, and water. Silicones are another type of mold release that are difficult to remove.

Various surface treatments can be utilized to make certain plastics paintable. Acid etching is one method which chemically oxidizes the surface. Flame treatment is another method used on inert plastics such as polyethylene.

Primers and base coats are used on plastics for various reasons. They may provide a better bond for the topcoat or fill in surface irregularities. Plastics can be electrostatically sprayed if a conductive prep coat is used. Plastic parts can be flow-coated or dipped.

Waterbone coatings are being applied to plastics with great success. However, the plastic substrate should not be hydroscopic, or water from the coating may be absorbed into the plastic and cause popping or adhesion problems. Generally, waterborne coatings are only slightly alkaline or acid so that they will not cause pH damage to the plastic.

AUTOMOBILES AND TRUCKS

Automobiles in the United States have been finished with lacquer and enamel exterior topcoats. General Motors has used lacquers that are acrylic polymers blended with plasticizers and with cellulosic polymers. Enamels are used for all other car lines and for all truck finishes. These are generally acrylic polymers blended with melamine resin at about a 70:30 ratio. These enamels are thermosetting and must be baked. Interior coatings are of the same composition as the exterior but with a reduced gloss.

Spray primers are used under the topcoat to provide properties of adhesion, appearance, metal filling, and so on. Many automobile parts are coated by dip priming or flow-coat methods. Automobile bodies are primed by electrodeposition. Table 15.4 shows the amount of paint used on a conventional

TABLE 15.4. PAINT USAGE PER CAR (FULL-SIZED) IN GALLONS.

	Lacquer	Lacquer	Enamel	Enamel
Topcoat:				
Lacquer (33% NVM)	2.4	2.4	—	—
Enamel (43% NVM)	—	—	2.1	2.1
Interior	0.15	0.15	0.15	0.15
Primer:				
Spray primer (55% NVM)	1.1	0.5	1.1	0.5
Dip primer (47% NVM)	0.3	—	0.3	—
Electrodeposition (38% NVM)	—	1.4	—	1.4
Underbody (42% NVM)	0.35	—	0.35	—
Miscellaneous	0.50	0.25	0.5	0.25
Total gallons:	4.80	4.70	4.50	4.40

full-sized automobile. This applies to solvent coatings. A total of 4.8 gallons of paint is used per car, using a lacquer system of which about 25 pounds is solvent. In an enamel system 4.5 gallons of paint is used, of which about 21 pounds is solvent.

Waterborne enamels are now being used on two automobile production lines in California. They are thermoset acrylic enamels. These coatings receive a 38-minute bake at 325°F of which the first 8 minutes is an induction period below 180°F to eliminate "popping."

Waterborne spray primers are usually epoxy modified water-dispersed materials with conventional pigments.

Waterborne dip primers are used on many automotive parts and are usually about 42–45% solids. A typical formular is shown in Table 15.5. Using waterborne paints on a conventional automobile will reduce the amount of solvent used to about 9 pounds per vehicle.

COIL COATING

Coil coating is a high-speed painting process in which coils of aluminum or steel strip are cleaned, treated, and coated with a baking enamel.

The coated metal is then formed into the finished product, which may be exterior siding, gutters, awnings, shelving, roof decking, highway guard rails, and so on. This process eliminates the need for the metal fabricator to perform slow and costly finishing operations after the products have been shaped and assembled.

With a coil coating line, the entire series of finishing operations is consolidated into one continuous process in one area. A roll of metal strip is successively unwound, straightened, cleaned and treated with a conversion coating, dried, roll-coated, baked to cure the coating, cooled, and rewound. Production lines coat metals from 1 inch to 72 inches wide and up to 16-gauge thickness, at line speeds of 200 to 500 feet per minute.

Prepainting the metal strip prior to fabrication produces more uniform paint thickness. On the coating line the metal is immediately painted after cleaning, a step that eliminates the possibility of surface contamination. Paint waste is reduced so that the cost of coating coil stock may be less than one-half that of spraying. The only loss of paint is from trimming and scraps from the fabrication operation.

The major advantage that continuous coil coating offers is its high production capacity. For example, a 60-inch-wide coater operating at an average speed of 200 feet per minute is capable of producing 72,000 square feet of coated strip per hour. Also, by the same token, a large amont of off-grade material can be produced in a short time so that all materials and steps in the operation must be closely controlled.

TABLE 15.5. DIPPING PRIMER (BLACK).

	Pounds	Gallons
Alkyd solution[a]	142.4	16.47
Potassium tripolyphosphate	1.1	.05
Furnace black	10.6	.71
Barytes	159.5	4.36
Micronized talc	42.5	1.89
Barium metaborate	53.2	1.91
Triethylamine	0.5	.09
Water	159.5	19.16
Pebble mill, 18 hours		
Alkyd Solution[a]	212.6	24.58
6% Manganese napthenate	0.6	.09
Butanol	5.3	.79
Water	248.7	29.90
	1056.5	100.00
42.9% NVM		
1.5:1.0 Pigment/binder by weight		
40–60 sec Viscosity #4 Ford Cup		
8.0–8.8 pH		
Weight/gallon: 10.57 lb		

[a]Arolon 363 WA850— Butoxy ethanol 25.0%
Triethylamine 3.3%
Water 71.7%

Many fabricators now purchase coated coil stock so that they can eliminate the painting process within their plant.

Coating Requirements

Film flexibility is the primary requirement for paints to be used in coil coatings. The coated metal may have to undergo forming, embossing, drilling, cutting, drawing, and stamping during fabrication.

Since a large amount of coil stock is used for exterior products, durability is a prime requirement for many applications. Exterior durability includes resistance to fading and chalking, gloss retention, resistance to erosion, and resistance to dirt pickup. In the case of residential and industrial siding the manufacturer usually gives a 20-year performance warranty backed by the coatings producers.

Coated stock must be able to stand rough handling without damage, including resistance to normal wear of the finished item.

Resin Systems

There are many types of resins used in coil coating systems. Important properties are cost, adhesion, flexibility, adaptability to embossing, and durability. These resin systems are rated in Table 15.6. Alkyd-melamines, which were the first system to be used on coil coating lines, are low in cost and give good performance for many applications. Acrylics give excellent results—and note how the durability is improved by using an emulsion over the solution type, because of the increase in the molecular weight of resin.

Table 15.7 shows the erosion rates of various coating systems on exposure. These are 1-mil films. Note that bakes coil coating systems have much lower erosion rates than coatings such as oil-base house paints and acrylic emulsion house paints, which are applied on site.

Tables 15.8 and 15.9 give formulas for an acrylic white coil coating enamel and a polyester white coil coating enamel. A formula for a clear baking varnish for direct roll-coater application is shown in Table 15.10.

BOARD PRODUCTS

The term "board products" includes hardboard, particle board, plywood, and overlay plywood. All have been important building products for a number of years, and with the increasing scarcity and increasing cost of quality lumber, their use has grown greatly in the past decade. In the past, wood products were sold in an unfinished condition for painting on the job site. However, factory-applied finishing is lower in cost and produces a better finish under controlled factory conditions.

TABLE 15.6. COIL COATING SYSTEMS.

Resin type	Cost	Adhesion	Flexibility	Adaptability to embossing	Durability
Alkyd–melamine	Low	G	F	P	G
Vinyl—organosol	Low	G	E	G	G
Vinyl—solution	Med.–high	E	E	G	F
Vinyl—alkyds	Med.	G	G	F	F
Polyesters	Med.	G	G	F	G
Acrylics—solution	Med.–high	E	E	G	F
Acrylic—emulsion	Med.	E	E	E	E
Silicone—polyesters	High	G	F	P	E
Silicone—acrylics	High	G	F	P	E
Fluorocarbons	Very high	G	E	E	E

E = excellent; G = good, F = fair; P = poor.

TABLE 15.7. WEIGHT LOSS BY EROSION ON EXPOSURE OF WHITE COATINGS.
(1-mil coatings, 2 years)

	%
Oil-base house paint	41.5
Acrylic emulsion house paint	30.2
Amine alkyd coil coating	10.2
Vinyl solution coil coating	4.0
Acrylic solution thermosetting coil coating	4.6
Vinyl organosol coil coating	6.0
Acrylic emulsion coil coating	7.2
Fluorocarbon organosol coil coating	0.5

TABLE 15.8. ACRYLIC COIL COATING ENAMEL (WHITE).

		Pounds	Gallons
Deionized water		46.4	5.57
N,N-Dimethylethanolamine		0.1	0.01
Ethylene glycol		3.4	0.33
Nonionic surfactant[a]		2.2	0.26
Dispersant[b]		7.3	0.84
Defoamer[c]		0.5	0.06
Titanium dioxide		211.7	6.14
Grind in a Cowles dissolver			
Defoamer[c]		271.7	13.21
Deionized water		35.4	4.25
N,N-Dimethylethanolamine		5.6	0.75
Acrylic-styrene latex[d] (43%)		543.4	61.89
Deionized water		97.7	11.50
Butyl carbitol		39.8	5.01
Melamine resin[e]		31.0	3.10
		1026.9	100.0
Viscosity—Brookfield 60 rpm	500 cp	Baking	1 min. 420°F
pH	90	Pencil hardness	2H
NVM	44.5%	Gloss—60°	88
		20°	55
		Impact—Gardner	
		Face	40 in lb
		Reverse	25 in lb

[a]Igepol CA 630—GAF Corporation.
[b]Tamol 731—Rohm & Haas Co.
[c]Foam Master VF—Diamond Shamrock Chemical Co.
[d]Ucar 4510—Union Carbide Corp.
[e]Cymel 303—American Cyanamid

TABLE 15.9. POLYESTER COIL COATING ENAMEL (WHITE).

	Pounds	Gallons
Titanium dioxide	282.1	8.47
Polyester resin[a] (70% NVM)	95.9	10.65
Water	145.6	17.51
Pebble mill 18 to 24 hr		
Polyester Resin[a] (70% NVM)	265.3	29.47
Hexamethoxy methyl melamine[b]	64.8	6.48
Trimethyl propanediol isobutyrate[c]	63.7	8.06
Dimethylethanolamine	1.7	.22
Water	159.1	19.14
	1078.2	100.0

NVM	55.6%
Viscosity	80–100 sec Ford Cup #4
Cure	1 min. 400°F
Pencil hardness	3H

[a]Arolon 465.WA8.70—Ashland Chemical Co.
[b]Cymel 301—American Cyanamid Co.
[c]Texanol—Eastman Chemical Co.

TABLE 15.10. CLEAR BAKING VARNISH FOR DIRECT ROLL-COATER APPLICATION.

	Pounds	Gallons
Acrylic latex[a] (43%)	701.9	80.0
Deionized water	27.5	3.3
N,N-Dimethylethanolamine	5.9	0.8
Hexylene glycol	88.4	11.5
Melamine resin[b]	44.2	4.4
Defoamer[c] (use if needed)	—	—
	867.9	100.0

NVM (33.6%)
Viscosity (Brookfield) LVT 60 rpm
Bake 4 min. at 360°F
Film properties on Bonderite 100 at 1 mil dry film thickness

Gardner gloss, 20°	100
Wedge bend	Passes
Pencil hardness	2H
Water resistance 45 min.	Excellent

[a]Ucar 4510—Union Carbide Co.
[b]Cymel 350—American Cyanamid Co.
[c]Foamaster VF—Diamond Shamrock Co.

Board-product finishes are applied by curtain, flow, or direct or reverse roller coatings or by spray at relatively high line speeds and low temperature cures. They must develop fast "block resistance" so that they can be stacked after curing without sticking.

The following is a list of coats used on board products:

Sealer coat	Provides uniform sizing and adhesion.
Sanding filler	Gives smooth, uniform surfaces by filling indentations, pores, and other imperfections present in the substrate.
Base coat	Provides a uniform white or colored base coat with good hiding power and good holdout for the topcoat.
Grain or print coat	Produces a simulated wood pattern or a decorative appearance on the finished board.
Clear or pigmented topcoat	Provides a protective coat with good resistance properties, and controls gloss.

Not all factory prefinished wood products will have the complete series of coatings applied.

Tables 15.11 and 15.12 give formulas for a clear sealer and a sanding filler for board coating.

CONTAINER COATINGS

Container coatings are applied to food, beer, beverage, and other cans, as well as closures, drums, and pails. Waterborne coatings have been used for the white base coat and the overvarnish. These coatings are usually applied by roller coat. Waterborne coatings have been used for the interior spray coat on

TABLE 15.11. CLEAR SEALER FOR BOARD COATING.

	Pounds	Gallons
Acrylic latex[a] (46.5)	181.1	20.3
Water	649.5	77.9
Wetting agent[b]	0.1	—
Butyl cellosolve[c]	13.7	1.8
	844.4	100.0
NVM—9.9%		

[a]Rhoplex AC-73—Rohm & Haas.
[b]Triton GR-7M—Rohm & Haas.
[c]Butyl cellosolve—Union Carbide.

TABLE 15.12. SANDING FILLER FOR BOARD COATING.

		Pounds	Gallons
Dispersant[a]		18.8	2.0
Dispersant[b]		3.9	0.5
Defoamer[c]		3.9	0.5
Water		96.8	11.6
Hydroxyethyl cellulose[d] (2.5%)		135.5	16.3
Barytes[e] (30-micron)		193.0	5.3
		290.3	12.9
Grind in high speed disperser for 12 min.			
Acrylic latex (46.5%)[f]		387.0	43.5
Butyl cellosolve		58.1	7.4
		1187.3	100.0
NVM	60.0%		
Pigment content by weight	50.0%		
pH	9.2		
Viscosity, K.U.	85		

[a]Tamol 850 (30%)—Rohm & Haas.
[b]Triton CF-10—Rohm & Haas.
[c]Nopco NXZ—Diamond Shamrock.
[d]Natrosol MR—Hercules, Inc.
[e]Duramite—Thompson-Weinman & Co.
[f]Rhoplex AC-73—Rohm & Haas.

two- and three-piece cans as well as the three-piece can side-seam spray. The end-sealing compound can be a waterborne product. One new method of coating two-piece cans is to apply a waterborne base coat on the entire can during the final stages of the cleaning operation and then bake the coating in an oven. Another experimental method is to coat the interior of the two-piece cans by filling them with a waterborne coating, applying a charge, and electrodepositing the coating on the inside of the can.

Interior coatings in contact with foods must meet rigid FDA regulations. Many of the coatings are applied to flat metal, which is then fabricated. These coatings must take shear, bending, and stretching operations.

Aqueous coatings are being applied to steel and aluminum for containers. Resistance properties, overbake color retention, adhesion, and coil coatability properties are all excellent.

APPLIANCE FINISHES

The term "appliance" covers a broad spectrum of products, ranging from small hand units to large commercial freezers. The appliances that constitute the major portion of the field are air conditioners, dishwashers, freezers, refrigerators, washers, and dryers.

Each class of appliance has certain definite requirements. Air conditioners, generally of the window type, are placed in a partially exterior environment and must therefore have good exterior durability and possess some degree of gloss retention. Freezers and refrigerators must have good resistance to staining from foods that will be stored in them. Dishwashers, washers, and dryers must have good resistance to detergents and bleaches. For all appliances, the desired qualities are basically a glossy, hard, tough, durable finish with good visual appearance.

ELECTROCOATING

Electrodeposition of paint is a process by which charged paint particles are electrically plated out of water suspension to coat a conductive object. The process is known as electrocoating, E. coating, elpo, electrodeposition, or electrophoretic painting.

Electrodeposition of an organic coating basically is a dipping process in which the coating is deposited by means of an electric current flowing between the part to be coated and the tank itself or separate electrodes adjacent to its inside surface. The part to be coated is either the anode (positive potential) or the cathode and the tank as the opposite potential. Because of the nature of the process the coating materials are restricted to waterborne paints of very low solids content.

The basic coating process is generally believed to consist of three simultaneous reactions: electrophoresis, electro-osmosis, and electrolysis.

Electrophoresis is defined as the movement of colloidol particles, dispersed in a liquid medium under the influence of an electric field. Negatively charged particles will migrate to the anode and be discharged, while positively charged particles migrate to the cathode.

Electro-osmosis is the movement of the liquid phase under the influence of an electric field. This is often referred to as the reverse of electrophoresis. It is as if colloidal particles are held against a membrane and unable to move under the influence of an electrical field. The medium in which they are dispersed will move in the opposite direction. It is this phenomenon that causes the deposited film to be relatively free of water when removed from the batch; in effect the water has been squeezed out, leaving a film with as little as 5% water.

The third reaction, electrolysis, is the dissociation and movement of ions under the influence of an electric field. Electrolysis is a more or less undesirable side effect, since it can cause difficulties such as a rough film due to gassing at the anode. It is a combination of the other two reactions that makes the process work.

Paint particles migrate to the part to be coated when a sufficient potential difference, usually between 80 and 180 volts, is applied between the part and the tank and/or separate electrodes. Thickness of the coating depends on several variables, including complexity of the part, time, temperature, voltage, solids content, and throwing power of the paint—that is, its ability to penetrate recessed areas. Coating thicknesses up to 1.5 mils are obtainable commercially and over 3.0 mils experimentally. The coating has a solids content of about 95% and is essentially water-insoluble.

The relatively high electrical resistance of the coating tends to reduce drastically the rate of deposition as thickness increases. The thickness on easily reaches portions of a part builds up rapidly to a near-maximum value and then levels off, while the thinner coating at recessed portions continues to increase in thickness until a very uniform paint film is deposited even on most complex parts. As with electroplating, if the recesses are too deep, the time required to produce a uniform coating of sufficient thickness may be too long to be practical. With such parts, electrocoating can be speeded up by using supplementary electrodes.

Advantages

Electrodeposition can provide several important advantages when painting metal parts; for example:

1. The process lends itself to total automation, reducing labor costs.
2. Intermixed parts with different configurations and sizes can be coated.
3. A more uniform coating thickness is obtained than with any other commercial painting process.
4. Good edge coverage (without heavy edges) and excellent coating of recesses produce better corrosion resistance.
5. Absence of runs and sags minimizes sanding time.
6. Significant paint savings are obtainable—as much as 30%.
7. There is no fire hazard because only waterborne paints are used.
8. Energy savings of as much as 30% are possible, comparing a low-bake electrocoating formula with a conventional enamel.

Disadvantages

Cost of equipment is higher than that of other coating processes, and there are these other disadvantages:

1. The coating material has to be fairly closely controlled as to temperature, pH, and alkali content to avoid paint spoilage.
2. Surface defects in the substrate are visible through the applied coatings.
3. Pretreatment and rinsing requirements are more stringent than with conventional dipping.
4. Need for additional electrodes for deep recesses may increase labor costs significantly.

5. Only a single film may be applied, and only on a metal substrate.
6. It is difficult and expensive to change colors.

Equipment

Pretreatment System: Thorough pretreatment is necessary prior to electrodeposition to achieve good corrosion resistance. The pretreatment system should include rust removal, degreasing, phosphating, and washing. Thorough rinsing of the pretreated parts is important to avoid dragging salts or acids into the bath.

Coating Tanks: A conventional sheet metal tank can be used, and in some cases conventional dipping tanks can be converted for electrocoating. In a continuous-coating installation, the tank is relatively narrow and long and has a slope of 45° at either end. For intermittent operation the tank can be much smaller.

Power Supply and the Electrical System: The power supply is conventional; its basic function is to supply dc power of sufficient voltage and current to coat parts at the production rates and with the coating material designated by the customer. The most frequently used method of supplying the dc power is to apply a positive voltage through a busbar and brushes to the part being coated, and to ground the tank which serves as a cathode.

Circulation: Because of the low solids content of the bath, good circulation is mandatory. Agitators are used in some equipment. Pumps are used in many installations to provide circulation.

Temperature: Typical bath temperatures for electrodeposition are between 80 and 100°F. Temperature is controlled by water-cooled heat exchangers and electric heaters. Normally, heat is generated during the operation of the electrodeposition tank.

Filtration: In order to maintain good film appearance as well as to avoid serious contamination of the tank, good filtration is mandatory for commercial installations. A magnetic filter is used to pick up metal particles dragged in with the parts.

Controls: In addition to controlling current and voltage, it is necessary to control the temperature of the batch, its conductivity, pH, alkaline content, and solids content. Solids content is measured about once a day and pH several times a day.

History

Electrophoretic migration and electrodeposition of colloidal particles were observed as early as 1809 and are described in the literature. Practical work on industrial applications can be considered to have started with anodic eletrodeposition of rubber latex in the United States in the period 1923–31.

Electrodeposition of resins was introduced in the United Kingdom in the 1930s by Crossee and Blackwell Limited, who were granted patents for the internal lacquering of food containers. One of the first commercial applications of electrodeposition of coatings was on an assembly-line basis in England in 1962. The Crowley Works of Pressed Steel Company began coating small parts.

In 1961, Ford Motor Company, after a five-year study of the process, set up a pilot line to translate the theory to practical application by coating wheels. In 1963 Ford began applying primers to automobile bodies.

Thus, the development of electropainting was carried out simultaneously on both sides of the Atlantic. This has resulted in two different approaches to making the paint and controlling alkalinity. In the Ford process, an amine such as ammonia is used to solubilize the resin. As the paint is deposited, amines are released in the tank, and the alkalinity increases. Acidic fresh paint is added to correct this change, to maintain solids and pH. In the Ford process the tank is used as the cathode, and the object to be coated is the anode.

In Europe in the electrocoat process used by Imperial Chemical Industries and the Pressed Steel Company, a strong alkali, such as potassium hydroxide, is used to solubilize the resin. Also, separate cathodes are employed instead of using the tank. The cathode is contained in a cathode box, one face of which consists of an ion exchange membrane. These membranes have the property of allowing positively charge ions to pass through on their way to the cathode. Then the excess alkali around the cathode is flushed away. This makes for simpler pH control.

Potassium hydroxide is cheaper than amines. However, dialysis cells must be used to contain the cathode, and they require maintenance and replacement. Also the corrosion resistance of the potassium-hydroxide-solubilized resins is often lower than that of their amine counterparts because some potassium ions are trapped in the film, a situation that leads to greater conductivity with an increased tendency toward corrosion.

Cost

A large-scale automotive operation including metal pretreatment and ovens, designated to hand 50 cars per hour, might cost \$3,000,000. This would require a tank 10 feet in width and 60 feet long, holding about 70,000 gallons of paint. Industrial installation of a 5,000-gallon tank might cost \$100,000.

Waterborne Coatings for Electrodeposition

The waterborne coatings formulations used in electrodeposition are of two types: emulsion and water-soluble. As a rule, the coatings are formulated as simply as possible, since resin and pigment must move in the same direction and be deposited at reproducible rates.

The anionic-type resins are high-acid-value resins rendered water-soluble or emulsifiable by neutralization with amines.

Cationic resins are based on nitrogen cations. They are simple analogs of the aniodic resins in which the carboxylate group,—COO^-, has been replaced with the amino group, $-N-H^+$.

A wide range of polymers are being used as primary resins in the electrodeposition process:

Maleinized drying oils
Styrenated and vinyl toluenated maleinized oil
Phenolic modified maleinized oils
Alkyds
Epoxy esters
Styrene allyl alcohol copolymer esters
Acrylic copolymers
Polyesters

In some cases a second resin such as a water-soluble melamine and certain latex emulsion polymers are included in the coating formulation to control thermoset properties or to adjust final film properties. These secondary resins usually are present at less than one-third the concentration of the primary resin.

Water-soluble coatings offer a number of advantages for electrodeposition, including good film continuity, simple formulas, good pigment binding, and good tank stability. However, there are some drawbacks. Only a few water-soluble materials are commercially available, and most air-drying-type water-soluble materials have poor water resistance. For this reason electrodeposited coatings based on water-soluble resins are of the baking type.

The most common water-soluble-type coatings used in electrodeposition are those containing carboxyl groups along the polymer chain. The polymer backbone may be based on a variety of resins, including alkyd and acrylic types. The carboxyl-containing polymers are not water-soluble in themselves, but are made water-soluble by reacting with a base such as potassium or sodium hydroxide, or an organic amine such as diethylamine. When solubilized resins of this type are under the influence of an electrical field, the positive ions (Na^+, NH_4^+—, etc.) migrate toward the cathode. There they discharge and are deposited. The negative ions of the polymer ($RCOO^-$) migrate toward and are deposited at the anode. Upon discharge at the anode the polymer may react with water and deposit as an acid or may react with a metal ion at or near the anode and deposit as the metal salt. During the deposition process the amine solubilizer may build up, which causes an increase in pH, a reduction in specific resistance, and a reduction in current usage.

There are a number of ways in which amine content and pH can be maintained at the desired level. One is to circulate the bath through an ion exchange column designed to remove basic ions. Another way is to use a coating that is deficient in amine.

Electrodeposited water-soluble coatings generally have high electrical resistance. They quickly coat all surfaces, and as all surfaces are coated, the current approaches zero so that the process stops for all practical purposes. Some coatings are 1 mil thick or less, while heavier coatings are as thick as 2 mils.

Emulsion Coatings

Emulsion coatings vehicles consist of particles of binder distributed in a water system. The particle size ranges from 0.1 to 10 microns.

One advantage of an emulsion-type coating is that it is possible to use high-molecular-weight polymers that will impart properties such as color retention, water and chemical resistance, and good mechanical strength. In addition, superior coatings are obtained on air drying.

One drawback of electrodeposited emulsion-type coatings is that they are usually complicated in makeup, containing pigments, latexes, thickeners, defoamers, and numerous other materials. Another drawback is that the mechanical, electrical, and chemical stability of an electrocoated emulsion may be insufficient for reliable production use, thus resulting in the need to dump the tank periodically.

Cathodic Systems

All of the early electrodeposition systems were anodic processes. However, recently many have been converted to cathodic processes.

The advantages of cathodic electrodeposition systems over anodic systems are as follows:

1. In cathode systems, unlike anodic systems, metal dissolution does not occur at the cathode. The absence of electro-dissolved metal in the resin film during cathode deposition results in better film properties, especially in the case of white electrodeposits over milled steel.
2. Hydrogen formed at the cathode may not react in an unfavorable way with the deposited resin film. This is not true for the oxygen generated at anodes, which tends to deteriorate film properties by oxidation.
3. Unlike the anodic process, cathodic electrocoating tends to deposit over contaminants in the metal surface so that they don't show up in the finished film.
4. Salt spray performance and humidity resistance are better with cathodic coatings.

5. Cathodic coatings have color consistency which is highly essential whenever welded areas need to be coated.

Converting an Anodic Tank to Cathodic

Conversion of tanks depends on the individual setup. If the tank is already coated and insulated, and has insulated electrodes on the side of the tank, conversion is fairly simple. The changeover consists basically of draining the tank and thoroughly cleaning out the system, replacing the black iron cathodes with flushable stainless steel anodes, and reversing the polarity of the rectifier.

If the tank is the cathode, the conversion is more complex and requires insulating the tank prior to making the electrode switch. Cleaning the anode system is of utmost importance, as the anodic system is alkaline and the cathodic system is acid so that if the two made contact a coagulated mess will occur.

Electrodeposition is used on many automotive parts such as frames, springs, housings, heat exchangers, wheels, and so on. It is also used for aluminum extrusion, electrical housings, tool chests and boxes, hospital carts, and so forth. The paint costs in an electrodeposition system are less than for a spray-applied solvent system. The cost is from 50 to 80% on an equal square footage–thickness comparison.

AUTODEPOSITION

Autodeposition is a process where good film performance is achieved without the use of an inorganic pretreatment. A dilute solution of latex reacts with the metallic surface under acidic conditions. There is an interaction between the ions removed from the surface and the latex. The basic process consists of six steps: cleaning, rinsing, coating, rinsing, reaction rinsing, and oven curing.

Autodeposition is a combination of an inorganic pretreatment like phosphating plus a prime coating like electrocoating. While it does replace these two processes, the coating is essentially an organic coating containing iron ions. The coating step occurs via a chemical oxidation–reduction reaction similar to those that occur in electroless plating of metals.

The electroless painting batch contains a 10% organic latex emulsion and pigment, plus hydrofluoric acid and an oxidant–hydrogen peroxide. The pH is 2.5 to 3.5. The acid initiates oxidation of the steel. Metallic iron is oxidized to ferric ion; the metal dissolves at the work surface where the ferric ion and the organic latex are precipitated from solution.

The coating reaction is not exothermic: neither the solution nor the part to be coated is heated. Nonferrous parts are not coated by the process.

The typical paint line will deposit an 0.8- to 1.0-mil film in 75 seconds. Bake is five minutes at a temperatuer of 275–300°F.

This process is used to coat automotive parts such as head-lamp housings frames. The coating is equal to a phosphated and electrocoated steel in salt spray, Olsen cup, cross-hatch, and knife tests.

Energy is saved by this process because (1) the only heated stage is the alkaline wash; (2) there is a reduced cure schedule, and oven requirements are lessened; (3) no electricity is required to deposit the coating.

16

Cement and Silicate Coatings

Two types of waterborne paints using inorganic binders are cement paints and silicate paints.

Inorganic binders have the advantage of being nonburnable, but lack flexibility and water resistance.

CEMENT PAINTS

Portland cement mixed with other ingredients acts as a paint binder when reacted with water. The paint is supplied as a powder to which water is added before use. Cement paints are used on rough surfaces such as concrete, masonry, and stucco. These paints dry to form hard, flat, porous films which permit water to pass through readily. Cement paints can be used on fresh masonry and are low in cost. The surface should be damp when they are applied and must be kept damp for several days to obtain proper curing. They should not be used in arid regions. When properly cured cement paints of good quality are quite durable, but when improperly cured they chalk excessively on exposure and then may present problems on repainting. Cement paints should not be used on cement–asbestos surfaces or over organic coatings. Cement paints are not suitable for surfaces such as concrete floors that are subjected to traffic.

Cement paints are composed chiefly of white portland cement together with some hydrated lime or siliceous aggregate or both. Cement–water paints are of two types (medium or high cement content) and two classes (with or without siliceous aggregate). The lime contributes easier brushability; the silica fills in rough surfaces. Calcium chloride may be added for its hydroscopic effect in drawing moisture from the air, which promotes proper curing and hardening. Titanium dioxide or zinc sulfide pigment may be added to improve wet opacity, and colored pigments added to produce tinted paints. About 1% calcium stearate may be incorporated to provide some degree of water repellency to the dried paint.

Composition of a typical cement paint is:

	Percent
White portland cement	75
Hydrated lime	17
Opaque pigment	3
Calcium chloride	4
Calcium or aluminum stearate	1
	100

Ingredients are mixed in a dry blender. Add 8 pounds of the above dry powder to one gallon of water.

Concrete and masonry surfaces that are to receive cement–water paints should be cleaned so that they are free of dirt, oil, grease, and efflorescence. Application of cement–water paints is best accomplished with a stiff fiber brush. To accomplish the necessary damp curing, the painted surface should be wet down with a fog spray 6 to 12 hours after application—a procedure that should be done three times a day for 48 hours after application of the final coat.

Properly applied and damp-cured, cement–water paints provide a durable coating for exposed concrete or masonry surfaces above or below grade, inside or outside.

SILICATE PAINTS

In silicate paints the binder is an alkali silicate. Silicates are nonpetroleum in origin, and silicone is readily available, as it makes up 16% of the earth's composition. Silicate paints will stand temperatures of up to 1000°F.

Silicate paints based on sodium silicate will resolubilize from atmospheric moisture and efflorescence on the surface. Ammonium, potassium, and lithium silicates are more water-resistant. The insolubility of lithium silicate glasses is attributable to the small radius of the lithium ion and its resultant high charge density.

The more completely the water is removed from the silicate, the greater will be its resistance to rehydration. Since the amount of moisture retained by a silicate film is governed primarily by the temperature to which it has been exposed, baking is often desirable. At the outset, the temperature should be increased slowly to 200–210°F to remove the excess water slowly. After much of the water is thus removed, curing can proceed at temperatures of 300–400°F. Sudden heating of the wet film should be avoided because it may form steam, which in turn will build enough pressure to cause blistering or puffing.

Where high temperatures cannot be used to insolubilize the film, permanency may sometimes be achieved by reaction with acid or solutions or heavy metal salts such as calcium, magnesium, or aluminum. Silicofluorides, zinc oxide, and whiting are commonly included to help produce insolubility. Reactive clays, too, under certain conditions are effective.

Sodium silicate films when dehydrated have good dielectric properties. The specific resistance at 20°C of a completely dehydrated sodium silicate is about 3×10^{10} ohm-centimeter, which is nearly the same as for window glass. Resistance decreases with increasing alkalinity and rising temperatures.

Silicate films per se, because of their brittleness, tend to fail by surface cracking and flaking. Modifiers can be used to improve adhesion and flexibility, and include such materials as bituminous resins, clays, and oil emulsions.

Alkali-resistant grades of pigments, such as are commonly used in latex paints, are satisfactory in sodium silicate systems. Calcium sulfate reacts with sodium silicate and, therefore, should be avoided as an inert pigment.

Thin films of silicate are applied to lithographic plates to render the surface hydrophilic and acceptable for further processing. Steel ingots have been coated with thin films of silicates to protect against corrosion. Silicate films protect silver flatwear against scratches during manufacture.

For some requirements, silicates are used alone as a grease-proofing agent for coating paper-carton stock. They may also be used as a priming coat to prevent paraffin, used for water resistance, from soaking excessively into the paper. Silicates have been used in combination with styrene-butadiene latex for exceptional protection to clay-filled board and sulfite-lined cylinder board against turpentine, cottenseed oil, and lard oil.

A thin coating of silicate* protects stainless steel, colored by controlled oxidation, to render it resistant to atmospheric conditions and abrasion, as well as cleaning agents commonly used on architectural or decorative finishes. The silicate film does not change the color of the coated steel, but rather adds depth of color to the steel sheet. Baking at about 480°F cures the silicate and substantially insolubilizes the film. The silicate coating is smooth, glossy, and resistant to scratching.

A coating containing vinyl chloride latex and sodium silicate has been used as a high-temperature coating to protect metal on rockets during firing.+

NASA has developed a zinc-rich coating using potassium silicate as the vehicle with excellent corrosion resistance:

Component	Parts by weight	*% Weight*
Potassium silicate solids 20K5.3[a]	64.5	21.0
Methyltrimethoxysilane	2.0	0.6
Zinc dust (3- to 50-micron)	240.0	78.4
	306.5	100.0

[a]20K5.3: The first number, 20, refers to the percent by weight of potassium silicate in the aqueous solution; the second, 5.3, refers to the ratio of silica to potassium oxide.

*U.S. Patent 3,125,471—Allegheny Ludlum Steel

+U.S. Patent 2,699,407—C. Martens, 1955

The substrates to which the coating is applied are generally metallic. For optimum results, the surface should first be cleaned with phosphoric acid or by sandblasting. Application may be by brush, but the most uniform finish is obtained by spraying.

Textile kilns are protected from corrosive attack by coating with silicate, then dusting when wet with portland cement to render the coating insoluble. Subsequent boiling with Epsom Salt solution gives an enamel-like surface.

Coatings of stainless steel powder in a silicate vehicle have been used for protective fire walls of aircraft and for coating the exhausts of JATO tubes. The dried film is said to weld to the base metal at 1700°F and does not flow, even at 2000°F.

The adhesion of fluorocarbon polymer coating to its substrate is improved using a composition with 10–75° lithium polysilicate, 25–90° of a fluorocarbon polymer, and a liquid carrier.*

*U.S. Patent 3,697,309—E. R. Werner—E. I Du Pont de Nemours & Co., 1972.

17

Miscellaneous Products

Included in this chapter are miscellaneous waterborne products. These products include aluminum paints, fire-retardant coatings, strippable coating, multicolored lacquers, asphalt coatings, driveway sealers, latex–cement compositions, and caulks.

ALUMINUM PAINTS

The usual solvent aluminum paint will gas and deleaf in the package in the presence of water. The water reacts with the aluminum powder, releasing hydrogen gas. For this reason, it has not been practical to make a waterborne aluminum paint.

However, if the aluminum paste is treated with diammonium hydrogen phosphate, a barrier coat is formed that renders the aluminum relatively passive.

The stabilized aluminum paste is produced as follows:

	Parts
Water	65
Diammonium hydrogen phosphate	1.5
Nonionic surfactant	0.5
Aluminum powder	35

The diammonium hydrogen phosphate and surfactant are dissolved in water and allowed to stand ten minutes; then the aluminum powder is added and mixed thoroughly, and aged 48 hours before being used in paint.

A typical aluminum paint is as follows:

	Pounds	*Gallons*
Polyvinyl acetate (homopolymer 58% solids)	610.0	65.6
Tricresyl phosphate	17.8	1.8
Aluminum paste	80.7	3.9
Carbitol solvent	17.8	2.1
Butyl carbitol acetate	17.8	2.2
Cellosize hydroxyethyl cellulose-300 (75% NVM)	40.4	4.7
Water	164.0	19.7
	948.5	100.0

The pH range of the system should be maintained in the range of 3.0 to 8.5. Aqueous aluminum coatings evolve hydrogen at a slow rate during storage and should, therefore, be stored in vented drums.

FIRE-RETARDANT COATINGS

There is considerable interest in intumescent or fire-retardant foaming coatings. These coatings, when applied to wood or other combustible surfaces, retard the spread of fire by foaming when subjected to heat, thus insulating the surface and retarding burning.

Most intumescent coatings now used have poor washability. This can be improved by the use of a washable topcoat, which makes a two-coat job. Most intumescent formulas require the use of a chemical combination of a polyol, a mono- or diammonium phosphate, and an amide.

On heating, these chemicals produce a gas that is trapped in the film causing it to foam. The gases liberated should not be exceedingly toxic. Chlorinated products can often produce phosgene, and proteins heated in an atmosphere with a deficiency of oxygen can produce deadly hydrogen cyanide gas.

In fire-retardant emulsion paints, polyvinyl acetate, polyvinyl chloride, and acrylic latices may be used. The latex must be stable in the presence of soluble inorganic salts.

The following is an effective intumescent latex paint:

	Parts
Dicyandiamide	10
Pentaerythritol	20
Monoammonium phosphate	56
Titanium dioxide	12
Plasticized polyvinyl chloride latex (50% NVM)	50

The intumescent agents and pigment are dispersed in water using high-shear mixing equipment. Then latex is added with mild stirring. Thickeners, defoamers, and so on, should be included in the formula. Certain pigments such as antimony oxide and borates add to the fire-resistant properties of the paint.

To test fire retardancy, a "wick and stick" test may be made. These tests use thin wooden sticks that can be coated with paint. After drying, the coated sticks are held over a burning wick for a fixed period of time, and the distance the flame has spread is measured. Some typical rates are shown below:

	Flame spread rate
Cement asbestos board	0
Red oak	100
Fir painted with intumescent coating	20–40

The best fire-retardant coatings produced today have a flame spread range of 10–15. However, these materials have little or no washability. Overcoating with a washable overcoat increases the flame spread rating to 20–35. Best results are obtained with an intumescent paint when coverages of 100 square feet per gallon are used. Since intumescent paints are hydroscopic, flexibility of the film varies with atmospheric humidity. Continued exposure to atmospheres of relatively high humidity reduces the intumescing property of the paint film.

MULTICOLORED LACQUERS

Multicolored paints produce an unusual decorative effect. Many of the multicolors used today are dispersions of several colored lacquers in a water medium. Spray application is used in most cases. An example of a multicolor lacquer formulation is as follows:

Part I

Fifty parts by weight of a 1% solution of methyl cellulose (15 cps) is placed in a mixing kettle provided with a two-bladed propeller agitator driven by an electric motor. The agitator is started, and while it is rotating at about 600 rpm, 100 parts of a lacquer having the following composition is poured slowly into the kettle:

Part I–White Nitrocellulose Lacquer

	Parts by weight
Titanium dioxide	12
Wet nitrocellulose (5–6 sec)	15
Ester gum	10
Castor oil	2
Dibutyl phthalate	2
Butyl alcohol	4
Butyl acetate	8
Methyl amyl acetate	13
Toluol	17
Xylol	17
	100

Agitation is continued for about five minutes after addition of the lacquer is completed.

Part II—Red Nitrocellulose Lacquer

A dispersion is prepared in the same manner as described above except that 12 parts of indian red oxide are used in place of 12 parts of titanium dioxide.

Parts I and II are then combined in a kettle and mixed with the agitator running at a speed of 500 rpm for five minutes. A multicolored lacquer produces a three-dimensional effect, covers in one coat, minimizes surface imperfections, and dries rapidly.

ASPHALT COATINGS

Because of their good adhesion, water resistance, chemical resistance, and durability, asphalt materials have a wide range of uses. They are used in the building and maintenance of roads, parking lots, playgrounds, tennis courts, and so on. Other applications are automobile undercoatings, corrosion-resistant coatings for metals, roofing, paper coating, and soil stabilization. Asphalts are available for use as hot melts and solvent cuts and in the form of emulsions. For many applications, emulsions are preferred because of the ease of application and reduced fire hazard.

Early asphalt emulsions were prepared by using crude soaps or clay as the emulsifier. These systems are similar to most paint emulsions as they are anionic. These emulsions rely on the evaporation of water and the absorption of water into the surface upon which it is being applied. In applications such as road building, wet weather will slow drying and hinder application. A typical anionic asphalt emulsion is:

	Parts
Asphalt bitumen	55
Water	44
Sodium linoleate-alkali	1
	100

In preparation, the water and soap are heated, and molten asphalt is added slowly with good agitation. Commercially, these emulsions are produced on a colloid mill. A stable emulsion is desired, but it should have a "fast break" so that on application it sets up quickly.

A recent development is cationic asphalt emulsions. These emulsions are produced using cationic amines as the emulsifying agent. Emulsions are produced in which the asphalt particles have a positive charge. Since most surfaces have a negative charge, the emulsion particles are attracted by the opposite charge, and deposition takes place immediately by breaking the emulsion. Surface and atmospheric conditions have little effect on this charge. Cationic emulsions set up very quickly, and water resistance is obtained much more quickly than with an anionic emulsion. This is one of the few applications where cationic emulsions are used for a commercial coating application. (See Figure 17.1.)

A typical cationic asphalt emulsion is as follows:

	Parts
Asphalt	65
Water	35
Tallow diamine	0.30
Calcium chloride	0.10
Glacial acetic acid	0.12

The emulsion is prepared by dissolving the acid and calcium chloride in water and then adding the diamine and agitating until complete solution takes place. The hot water solution (180°F) is then added slowly to molten asphalt (250°F) with good agitation.

There are many formulas for the preparation of asphalt or bituminous emulsions described in the patent literature. Most show the use of low-cost emulsifiers such as starch, tall oil, polymerized rosin, petroleum hydrocarbon insoluble rosin, rosin esters, ligno sulfonates, polymerized fatty acids, protein

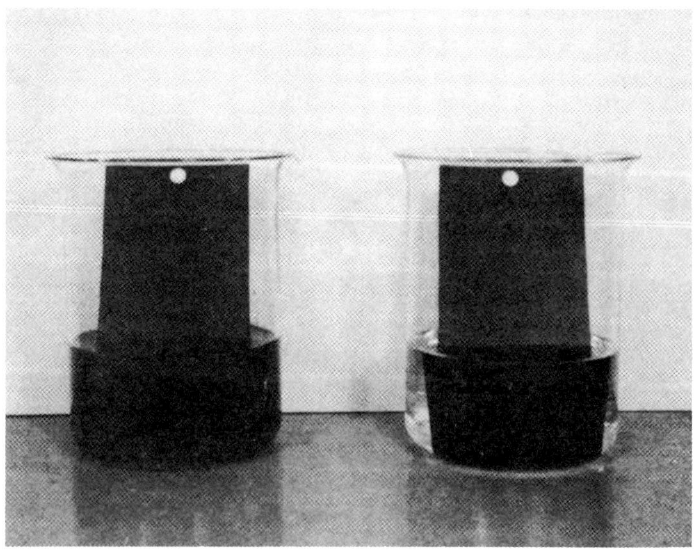

FIGURE 17.1. Rapid water resistance of cationic emulsions: metal panels coated with emulsion paints and air-dried two minutes before immersing in water: left, anionic; right, cationic.

meal, soya flour, and so on. Other systems use sodium silicate, amine emulsifiers, trisodium phosphate, and high-molecular-weight carboxylic acids. Many asphalt emulsions use clay as the stabilizer. The water resistance of asphalt emulsions can be improved by the addition of materials such as citric acid; ammonium chloride, sulfate, or nitrate; calcium, barium, magnesium, or strontium chloride; lead, titanium, zirconium, or tin salts; zinc, cadmium, or mercury salts; or fluorosilicates. The addition of 1 to 5% acrylonitrile-butadiene latex to an asphalt emulsion improves overall properties.

PORTLAND CEMENT COMPOSITIONS WITH LATEX

The addition of synthetic latices to portland cement produces a product with improved physical properties. Latices such as styrene-butadiene, polyvinyl acetate, and acrylic are used for this purpose in amounts of from 10 to 40% latex to cement on a solid basis. The tensile and compressive strength is improved particularly in a dry cure. The chemical resistance is improved. The bond strength is improved to the extent that latex–cement combinations are used for patching applications.

Fortified cements of this type are used for industrial floors because of their improved toughness and resistance to oils and greases. Stucco and plaster fortified with latex have improved impact resistance. Polyvinyl acetate is available in the form of a dried powder that, when added to water, will form an emulsion. It is, therefore, possible to make up powder paints and concrete mixes containing polyvinyl acetate resin.

STRIPPABLE COATINGS

The main purpose of strippable coatings is to protect finished surfaces during storage, shipping, and installation. Films of 1 to 2 mils often provide adequate protection against weathering. Where mechanical abuse is the primary consideration, a dry film thickness of 4 mils or more should be used.

A strippable coating should have toughness and block resistance combined with a high level of cohesion for good film strength to provide easy stripping at normal film thickness.

Strippable coatings are used on metal surfaces such as aluminum, steel, brass, and so forth; on plastic surfaces such as acrylics; and on polyethylene, glass, rubber, and painted and lacquered surfaces.

One use of strippable coatings is as a masking compound for acrylic signs. The sheet is coated with the strippable coating, letters are cut out, and this area is stripped out to act as a stencil. After spraying with paint the mask is stripped off. The strippable coating must show good adhesion where edges are cut so as to give sharp letters.

Strippable coatings are used to protect furniture and paneling from the factory to consumer.

Porous, absorbent, or rough surfaces give high levels of mechanical adhesion; so they are not suitable for strippable coatings.

A formula for a strippable coating is shown in Table 17.1.

DRIVEWAY SEALERS

Asphalts and macadams used for surfacing driveways and parking areas develop small cracks on the surface from exposure to the elements. Water freezing in the cracks will eventually cause break-up of the asphalt. Emulsified asphalt products are used as maintenance sealers on these surfaces. Acrylic sealers offer advantages of quick drying and durability.

A formula for a driveway sealer is shown in Table 17.2.

TABLE 17.1. STRIPPABLE COATINGS.

		Parts by weight	
Water		1.50	
Thickener[a]		0.75	
Coalescing agent[b]		1.50	
Soya lecithin[c]		1.00	
Antirust[d]		.25	
Premix above and then add:			
Acrylic latex[e] (45%)		95.00	
		100.00	
Solids	44.75%	Viscosity	75 cp
pH	8.7	Weight/gallon	8.50
Minimum filming temperature	65°F		

[a]Acrysol G-110—Rohm & Haas.
[b]Butyl Carbitol—Union Carbide Corp.
[c]551—Rowe & Rowe.
[d]Victawet 35B—Stauffer Chemical Co.
[e]Ucar 4312—Union Carbide Corp.

ACRYLIC LATEX CAULKS

Acrylic latex caulks are high-performance products with low shrinkage, initial low temperature flexibility, resistance to discoloration on aging or ultraviolet exposure, and dry and wet adhesion to alkyd paints, gloss, glazed ceramic, and concrete substrates. These caulks are produced from acrylic latices specifically designed for this use. They have low glass transition temperatures.

Table 17.3 presents an acrylic latex caulk formula and gives performance data for the caulk.

TABLE 17.2. SEMIGLOSS DRIVEWAY SEALER. (BLACK)

		Pounds	Gallons
Propylene glycol	premix	35.0	4.07
Hydroxyethyl cellulose[a]	premix	2.0	0.18
Surfactant[b]		4.3	0.47
Surfactant[c]		1.0	0.11
Antifoamer[d]		1.0	0.13
Preservative[e]		1.0	0.12
Water		120.0	14.40
Carbon black		20.0	2.50
Clay[f]		173.1	8.05
Grind above in high-speed mill			
Acrylic latex[g]		546.7	61.67
Coalescing agent[h]	premix	25.5	3.34
Antifoamer[d]	premix	1.0	0.13
Ammonium hydroxide	premix	2.0	0.27
Water		35.7	4.28
		968.3	99.72

Weight/gallon	9.71 lb	Pigment volume content	23%
Viscosity, K.U.	64–68	Volume solids	35%

[a]Natrosol 250 MR—Hercules, Inc.
[b]Tamol 731—Rohm & Haas.
[c]Triton CF-10—Rohm & Haas.
[d]Dapro 881-S—Daniel Products.
[e]Super Ad-It—Tenneco Chemicals, Inc.
[f]ASP-170—Engelhard.
[g]Rhoplex AC-61—Rohm & Haas.

TABLE 17.3. ACRYLIC LATEX CAULK.

		Pounds
Acrylic latex[a] (65%)		516.5
Biocide[b]		0.1
Wetting agent[c]		10.4
Ethylene glycol	premix	2.9
Methyl cellulose[d]		1.9
Dispersant[e]		6.3
Dispersant[f]		1.5
Calcium carbonate[g]		721.5
Titanium dioxide[h]		16.3
Mix the above materials for 1½ hr in a high-shear, low-speed mixer.		
Mineral spirits[i]		29.3
Silane adhesion promoter[j]		0.4
Add the above and mix 10 minutes more.		
Defoamer[k]		
Add the above and mix 5 minutes more.		
Pigment/binder ratio 2.2/1	100 gal.	1313.2

[a]Rhoplex LC-67—Rohm & Haas.
[b]Proxel CRL—Merck Chem. Div.
[c]Triton X405—Rohm & Haas.
[d]Methocel E4M—Dow Chem. Co.
[e]Calgon T—Calgon Corp.
[f]Tamol 850—Rohm & Haas.
[g]Camel Tex—H. T. Campbell Sons Corp.
[h]Ti-Pure R-901—E. I. du Pont de Nemours & Co.
[i]Varsul No. 1—Exxon.
[j]Silane Z.6040—Dow Corning Corp.
[k]Nopco NXZ—Nopco Chemical Div.

TABLE 17.3. A (continued)

Performance of the Caulk	Pounds
Consistency (seconds):	
Initial	9–12
After 30 days @ 50°C	20–30
After five freeze-thaw cycles, 18°C	35–55
Slump (inches):	
Channel method	0
Color:	
Initial	White
After 300 hr Fade-Ometer	No change
After 300 hr Weather-Ometer	No change
Tensile strength (psi):	
At maximum stress	15–20
At break	22–28
Elongation (%):	
At maximum stress	300–400
At break	450–550
Recovery @ 25% elongation (%)	58–63
Low temperature flexibility (15°F):	
¼″ slab on aluminum 180° bend	No cracking

18
Color in Paint

Color is a psychophysical reaction as seen by the eye and interpreted by the brain. The eye-to-brain communication is a science in itself, but we will be concerned only with how the normal process works for those people who have to make color decisions or judgments. They must have normal color vision. This is referred to as having satisfactory color aptitude. There are several recognized tests, such as the Federation of Societies of Paint Technology Color Aptitude Test, to check an observer's ability to see color difference. We are concerned here with color as related to paint finishes; so we will consider it in this manner. The essentials for visual observation of color are a light source, an object, and an observer.

VISIBLE SPECTRUM

We are aware of color as light when the retina of our eye is exposed to the action of electromagnetic radiation having a wavelength of between 400 and 700 mμ (millimicrons). The wavelength of the color we see ranges from violet (400–430 mμ), blue (430–485 mμ), green (485–570 mμ), yellow (570–585 mμ), orange (585–610mμ) to red (above 610 mμ). This range composes what is known as the visible spectrum of colors. (See Figure 18.1.)

This range of color can also be recorded and shown by the use of an instrument known as a spectrophotometer. The recorded range of color is known as a spectrophotometric curve. The curve shows the amount of light reflected from a sample at a wavelength from 400 to 700 mμ.

Figure 18.2 shows a spectrophotometric curve of a letdown of phthalocyanine blue in titanium dioxide. The sample has a high reflectance in the

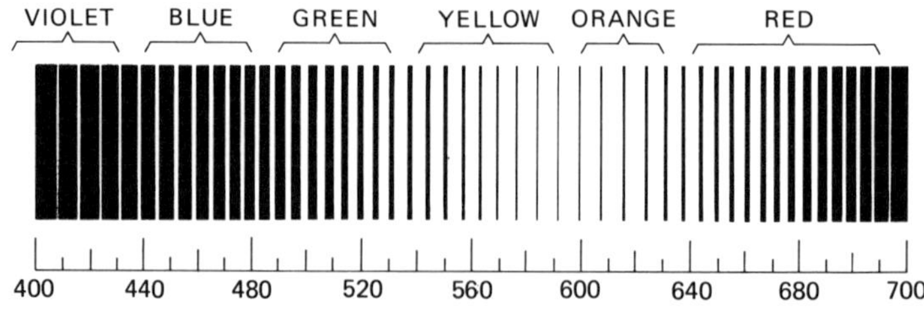

FIGURE 18.1. The visible spectrum.

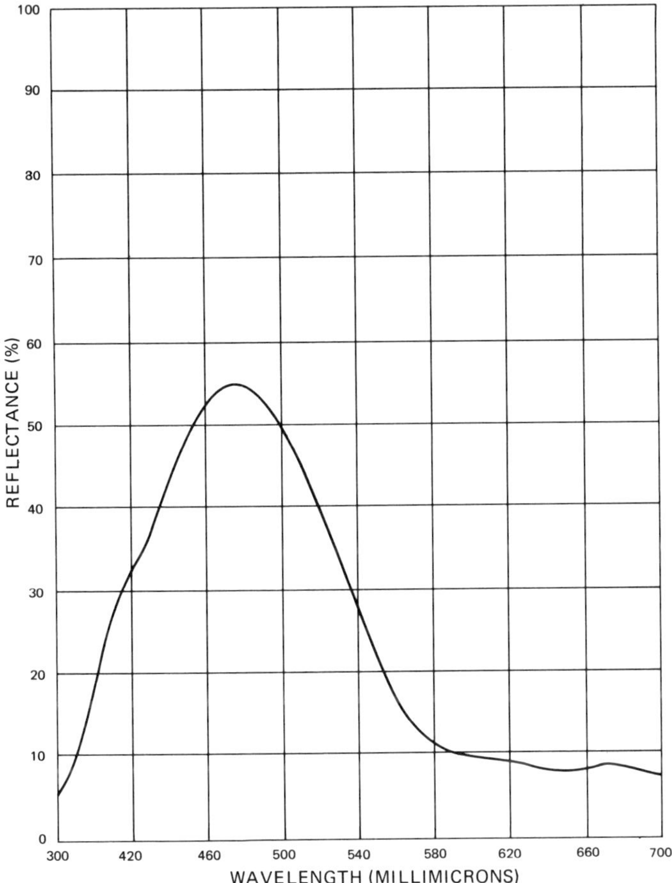

FIGURE 18.2. Spectrophotometric curve of letdown of pthalocyanine blue.

430–458 mμ region of the spectrum; therefore, it appears blue. Toluidine red has very little reflectance in any part of the spectrum except the red region; consequently it appears red (Figure 18.3). If a sample has a fairly equal distribution of light across the spectrum, it appears white (Figure 18.4) or various degrees of gray (Figure 18.5), depending upon the total light reflectance. Spectrophotometric curves have many useful functions in colorimetry.

THREE DIMENSIONS OF COLOR

Color cannot be described by a single word or characteristic. Color by its definition has dimension. To describe color adequately, all three dimensions must be characterized.

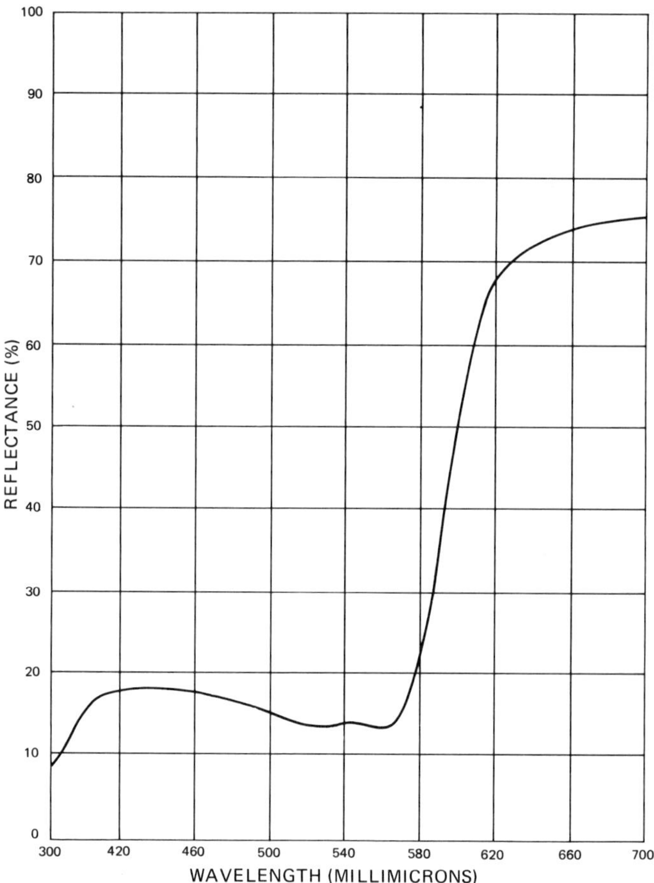

FIGURE 18.3. Spectrophotometric curve of letdown of toluidine red in TiO_2.

The first dimension of color is *reflectance,* which is also defined as value or lightness. The reflectance of a paint film is the amount of light that it reflects, regardless of its color. By this we mean that green, red, blue, and yellow can have the same light reflectance. Perfect white reflects 100% of the light that strikes it, and perfect black absorbs all of the light, reflecting none, so that it has zero reflectance. All other colors fall between white and black on this scale, as shown in Figure 18.6.

The second dimension of color is *hue,* which is the attribute of color that distinguishes green from red and blue from yellow. It is commonly referred to as the color of the object.

The third dimension of color is *saturation* or chroma. This dimension is the most difficult to understand and describe of the three dimensions. Saturation is the variable in color that denotes its departure from grayness. Let us assume we have a red and a gray that have the same reflectance. Now if we take four parts of red and one part of gray, we have a mixture of 80% red and 20% gray. The red will appear less brilliant or less red. It has not become less orange or blue but less red. It is now less saturated. If we make a mixture of 50% red and 50% gray, it will appear more gray, and so on with additional amounts of gray until the red is diluted to the point where the sample no longer appears red but is gray. The change in saturation has been obtained without any increase or decrease in reflectance.

We cannot assume that whenever a color is added to another color only one dimension will be affected at a time. Sometimes only two dimensions are affected, but usually all three are involved. If to the red we add 20% black instead of gray with the same reflectance, we will immediately dilute the red

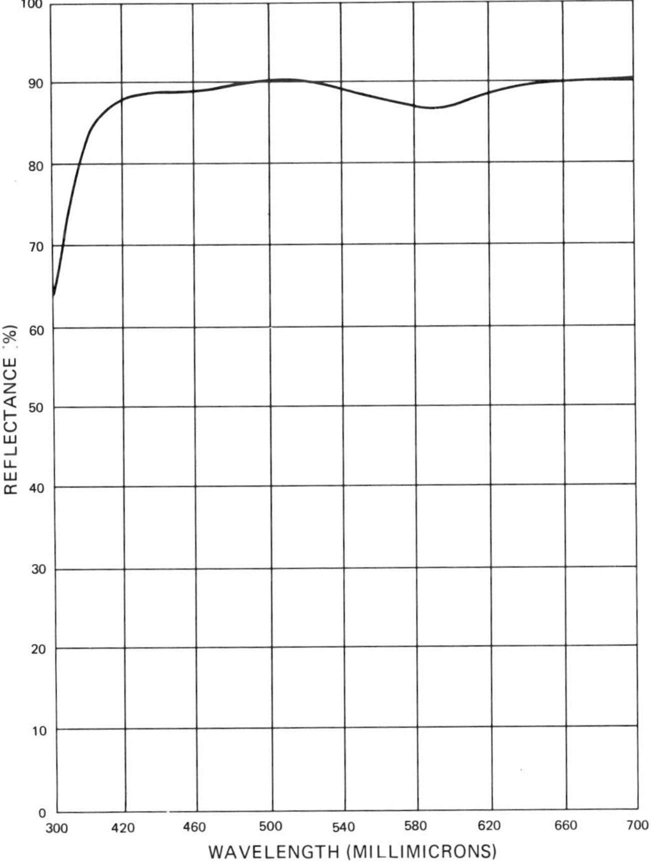

FIGURE 18.4. Spectrophotometric curve of a white enamel.

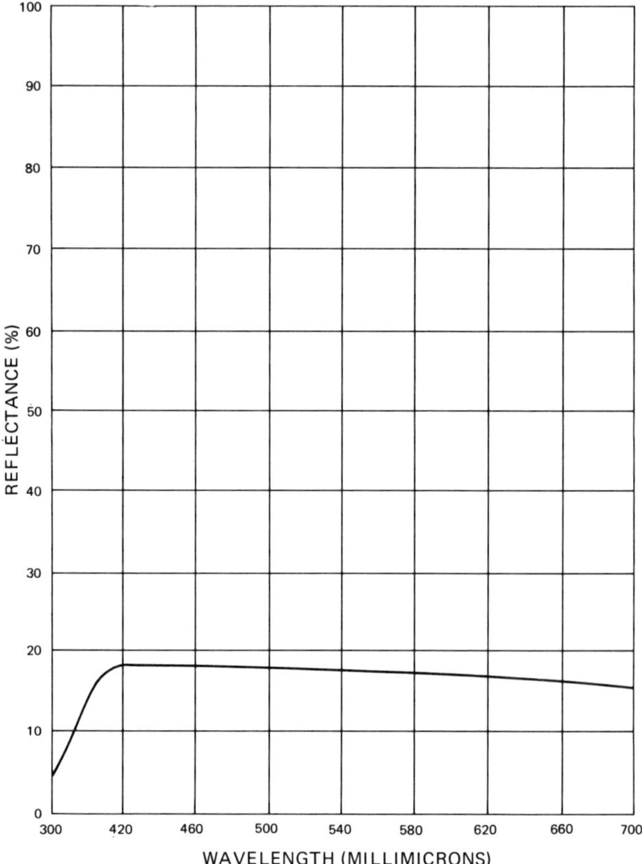

FIGURE 18.5. Spectrophotometric curve of medium-gray enamel.

by 20%. Therefore, it will move closer to the center of color space (Figure 18.6) because now our mixture is 80% red and 20% black, which is less saturated than 100% red. When we add a material that has lower reflectance than the red, the resulting mixture will also have a lower reflectance; so it will drop further down in color space. The sample will move down and in by the addition of black.

If we add white to the sample, the same thing will happen, only in the reverse direction. The red is now more diluted. In other words, it is not as brilliant or as saturated, and we have added material that has higher reflectance than the red. So now the color moves further up in color space and also into the center. White has the same effect on saturation as gray (i.e., moving it into the center of color space).

At this point the hue of our color has not been affected because all we have added to it has been white, gray, or black. Now let us take the red and add to it a yellow, which has the same degree of saturation and the same reflectance. Of course, yellow will have a different hue or dominant wavelength from red.

This mixture of two colors of different hue results in a third color, in this case orange. As long as the two colors that we mix have the same reflectance and the same saturation, the resulting color changes only in hue. However, if the yellow had a higher reflectance than the red and was less saturated, the mixture would then change in all three dimensions.

This is an example of what usually happens when colors are mixed in normal shading operations. When we view the results from mixtures of several colors, we mentally sum up all the steps that we have described and usually refer to it as a color change. However, color-measuring instruments resolve each step in a logical order and assign numerical values to them. For this reason we must understand the three dimensions of color: (1) lightness, (2) hue, and (3) saturation.

GLOSS AND ITS EFFECT ON COLOR

Gloss has such a profound effect on color that it is impossible to discuss one without the other. This is especially true of colors having low reflectance. A

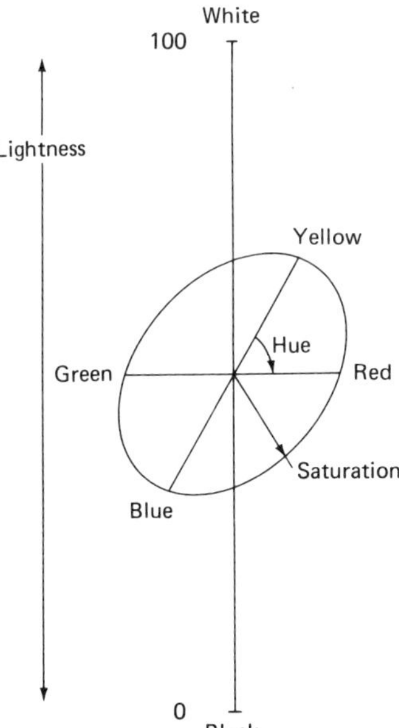

FIGURE 18.6. Relation of the three dimensions of color to each other. Reflectance is represented vertically, whereas the circular plane represents color space.

very slight change in gloss will have a noticeable effect on color. This can be shown by instrumental data as well as by visual observation. Samples with higher gloss appear deeper in color. (See Figure 18.7.)

LIGHT SOURCES FOR COLOR MATCHING

In daylight the complete range of visible waves is from 400 to 700 mμ. If this light strikes an opaque body, such as a painted surface, and all of the light is reflected, the surface appears white. The majority of surfaces that we deal with do not reflect all light that strikes them, but because of special optical properties, absorb a greater fraction of some wavelengths than others.

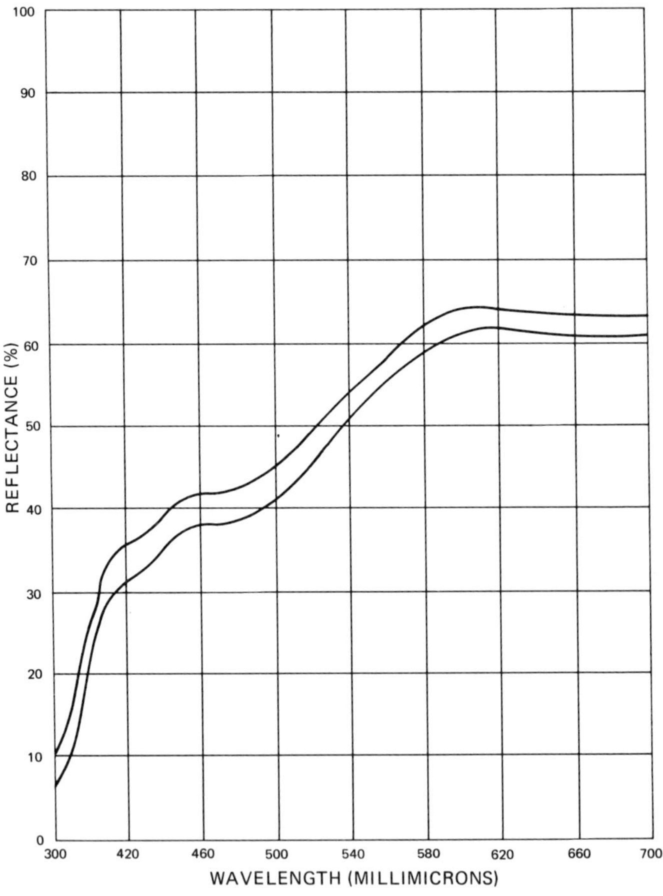

FIGURE 18.7. Effect of gloss on color. The two samples are the same color, differing only in gloss. The lower curve represents the sample with higher gloss. The upper curve represents the sample of lower gloss.

The reflected light thus lacks the wavelength region of high absorption, and the surface no longer appears white but has a definite color, depending upon the degree and spectral location of the absorbed energy. A red object absorbs all of the colors in the spectrum except red; that portion of the light is reflected back to the observer, and the object appears red. With a blue object, only the blue light comes back to the observer.

From this example, the importance of the light source in making color judgment is evident. If a light source contained no red or was deficient in red, then the samples viewed under it would not appear red, or as red as they should be, depending upon how deficient the light was in the red region.

North daylight has long been considered the most satisfactory source of light for color matching, having a fairly even distribution of energy across the complete visible spectrum. North daylight varies, depending upon the time of day. Since paint factories operate around the clock they must use artificial light sources for color matching. Artificial light sources have been developed to reproduce north daylight as closely as possible. Common artificial light sources are tungsten and fluorescent lamps. More detailed specifications for lighting conditions are given in ASTM D 1729 60 T "Visual Evaluation of Color Difference in Opaque Materials."

A metameric match is one in which the sample matches the standard under one illumination but does not match it under a different illumination. The spectrophotometric curves of the two samples will not match, and the curves will not be parallel.

COLOR FORMULATION

Requirements for a color formulation are as follows:

1. It must be able to match the color with the pigments selected.
2. It must produce matches that are not metameric to one another.
3. Performance requirements such as chemical resistance, lightfastness, and so on, must be met.
4. It must have the lowest possible cost.
5. Commonly stocked pigments should be used.
6. The hiding and tinting strength requirements of the product must be produced consistently.

The four-pigment concept of color formulation states that all colors can be matched with a minimum–maximum of four pigments with one held constant. The pigments must be selected so that the color to be matched lies within their gamut and gives adequate control of the three visual variables of color: hue, saturation, and lightness. There will usually be more than one set of four pigments that will match the color. Selection of the set will depend upon the performance requirements of the product, cost, and so forth.

Universal Colorants

The paint-buying public wants paints in a wide variety of colors and shades. A dealer may carry as many as 12 colors in a paint line. Many more colors are obtained by using a tint base line.

Tint base paints are white coatings that are tinted or colored in the store after the buyer has made his color selection. This coloring process is accomplished by the addition of a predetermined amount of colorant or several colorants to the tint base. The paint is then agitated on a paint shaker to disperse the color in the tint base.

Most paint stores have one series of tinting colors (about 8 to 12 different colorants) that are used in many lines of paints, including alkyd solvent paints and latex paints for interior and exterior applications. These are referred to as universal colorants because of their diversity of uses. With a tint base color system, as many as 1000 to 2000 colors may be offered in a single line.

The universal colorant must be of standard uniform strength from batch to batch so that when a metered amount of color is added to the paint, uniform color is produced from batch to batch.

A formula for a universal tinting color is given in Table 18.1.

Tinting colors may be produced in glycol, alkyd, or acrylic systems with surfactants, solvents, and suspending agents. With the outlawing of lead colorants, there are many pigments that should not be used. In some cases two tinting bases are available in a particular color, one for interior use and another for exterior use. In the case of yellow, the pigment for exterior use costs about three times more than the interior one.

The standard tint base white usually has two pounds of titanium dioxide per gallon. In some cases, a neutral tint base containing only inert pigments is also available for certain colors. The titanium dioxide contributes hiding power to colors in the critical yellow, orange, and light-green ranges. Also available are flatting bases, which can be added to enamel products to reduce the gloss.

TABLE 18.1. UNIVERSAL TINTING COLOR.

	Pounds
Lemon chrome yellow	415
China clay	225
Isoctyl phenoxypolyglycol ether	200
Mineral spirits	100
Hexylene glycol	250

The above ingredients are milled in a pebble mill to enamel fineness. The mill is emptied, and washed with 100 pounds of hexylene glycol. The yield is 100 gallons of tinting base with a weight per gallon of 12.7 lbs.

It is also necessary that the tint base show good acceptance of different quantities of colorants, as well as maintain its color once it has been tinted. Tint bases that do not have good color acceptance and good stability may cause the following problems:

1. White or color streaking and floating, which gives a nonuniform appearance.
2. Color difference between brushed and rolled areas.
3. Poor touchup.
4. Color drifting on aging in the container.
5. Poor tintability on aging because of loss of tintability, which gives a color that does not match the color chips.

A typical series of tinting colors might be:

Color	Pigment	% Pigment	Wt./Gal.
Black	Lamp black	20.0	9.10
Blue	Phthalo blue	20.0	9.00
Green	Phthalo green	25.0	9.80
Umber	Raw umber	40.0	11.70
Red	Toluidine red	35.0	9.35
Red	Iron oxide	60.0	16.05
Yellow	Diarylide yellow	25.0	8.90
Yellow	Iron oxide	50.0	13.30
Orange	Dinitranaline orange	35.0	9.50
Violet	Quinacridone violet	15.0	8.70

19

Testing of Raw Materials and Finished Goods

RAW MATERIALS

A modern paint plant maintains a raw materials testing laboratory to check incoming raw materials. Although all raw materials are purchased on specifications, a few spot tests are made to see if the supplier is furnishing raw materials within specifications. The more uniform the raw materials, the fewer the problems encountered in production and the more uniform the finished products. Testing also ensures that the performance of the paint in the field will be up to standard.

The modern raw materials laboratory uses both "wet chemistry" and physical testing (viscosities, melting points), as well as the latest instrumental methods to accomplish its goals.

An important consideration in the testing of raw materials is proper sampling to be sure that a representative sample is obtained. An approved test method should be used. The ASTM (American Standards Test Methods) have developed tests for evaluating raw materials as well as finished products.

The broad classification of raw materials that are tested is:

Pigments, inerts
Oils
Resins, polymers, plasticizers
Solvents
Miscellaneous—driers, biocides, surfactants, and so forth

Pigments

The general tests run on pigments are oil absorption, color, tinting strength, and particles retained by a 325-mesh screen (coarse particles). Special tests are run by instruments, such as emission spectrographic analysis for trace or contaminant analysis, X-ray diffraction to determine crystal structure, and infrared spectrometry for composition.

Oil absorption

This measurement is defined as the pounds of oil per 100 pounds of pigment to obtain a stiff puttylike mixture. It is not of much value in waterborne systems, as water absorption is more meaningful and is not always comparable to oil absorption.

Color

Color is judged by comparing pigment pastes of the sample vs. a standard. Color is now usually measured by instruments.

Tinting Strength

Tinting strength of colored pigments is measured by mulling pastes under controlled conditions with a standard white paste in the ratio of 1:20 or 1:50. If the tinting strength of the sample is not equal to that of the standard, the operation is repeated using smaller amounts of the stronger paste until the reduction of the sample paste appears equivalent in color strength to that of the standard paste. The tinting strength is calculated by taking the ratio of the weight of standard paste to that of sample paste used to obtain equality in strength. Multiplying this ratio by 100 will convert the result to percent.

Tinting strength of white pigments is measured by mixing specified amounts of white paste with colored pastes and comparing to the standard sample. If a sample is lighter in color than the reference standard, it has greater tinting strength.

Coarse Particles

This usually refers to the residue left on a 325-mesh screen (44-micron) after flushing a given weight with water. Instruments are available that can run a complete particle-size distribution down to 0.5 micron in about an hour's time.

Oils

Oils, while important in solvent paints, are of relatively minor importance in waterborne paints. Specific gravity, acid number, saponification number, iodine number, color, and viscosity are usually determined.

Resins

Water-soluble resins are usually tested for % NVM, acid number, color, weight/gallon, and viscosity.

The acid number is determined by titrating with an alkali and is used to measure the progress of the esterification of the resin during production. The acid value also determines the amount of amine needed to solubilize the resin.

Color is measured on a Gardner Hellige Comparator. The resin solution is placed in a Gardner bubble viscosity tube and compared to a series of standard colors ranging from 1 to 18 (the lower the number, the lighter the color).

Viscosity is measured by the use of Gardner-Holdt viscosity tubes, which compare the time of an air bubble to rise in a vertical tube to that of a standard tube. The tubes are designated in letters from A5 to Z10. Approximate viscosity in poises is shown in Table 19.1.

The NVM (nonvolatile matter) or solids content is determined by evaporating the solvent under standard conditions and weighing the residue.

For latex materials, slightly different tests are used: NVM, pH, weight/gallon, viscosity, particle size, and MFT.

The minimum film-forming temperature (MFT) of a latex is the minimum temperature at which a latex will form a continuous film. When a polymer latex is spread over a substrate and allowed to evaporate, the residual polymer coalesces into a continuous film if the temperature is above a critical value, the minimum film-forming temperature (MFT). The value of MFT is usually

TABLE 19.1. GARDNER-HOLDT VISCOSITY.

Gardner-Holdt bubble tube	Poises	Gardner-Holdt bubble tube	Poises
A5	0.00505	P	4.00
A4	0.0624	Q	4.35
A3	0.144	R	4.70
A2	0.220	S	5.0
A1	0.321	T	5.5
A	0.50	U	6.27
B	0.65	V	8.84
C	0.85	W	10.7
D	1.00	X	12.9
E	1.25	Y	17.6
F	1.40	Z	22.7
G	1.65	Z1	27.0
H	2.00	Z2	36.2
I	2.25	Z3	46.3
J	2.30	Z4	63.4
K	2.75	Z5	98.5
L	3.00	Z6	148
M	3.20	Z7	388
N	3.40	Z8	590
O	3.70	Z9	855
		Z10	1066

close to the gloss transition temperature (Tg), though the porosity of the substrate may affect the level of MFT. Below MFT, powdery, weak compositions are obtained. A laboratory instrument has been developed for rapid, reproducible measurements of MFT between temperatures 0 and 100°C. Means for control of humidity over the film can be provided to permit the study of atmospheric effects on film formation.

Solvents

Organic solvents specifications usually give specific gravity, distillation range, flash point, solubility parameter, and evaporation rate (butyl acetate = 1).

Distillation range characterizes the solvent more fully than any other test, since it characterizes the solvent with respect to the drying rate it will impart to the coating in which it is used.

PAINT PRODUCTS

Paint products are tested for properties that show that the paint has been made according to formula (weight/gallon, etc.), that it will apply satisfactorily (color, viscosity, drying, etc.), and that it will perform adequately in the field (salt spray, exterior durability, etc). Tests such as weight/gallon, color, viscosity, and drying may be run on each batch, whereas performance tests may be run only periodically.

Properties and *test methods* specific to waterborne paints can be classified into the following categories:

I. Liquid Paint Properties
   A. Package stability, appearance, homogeneity
      1. Freeze, thaw cycles with latex paints
      2. Hot room accelerated coagulation
      3. pH and specific resistance stability
   B. Rheology changes with temperature, dilutions, shear rates, micelle formation, phase inversion
   C. Biodeterioration

II. Film Properties
   A. Low-temperature coalescence
   B. Humidity effects on drying
   C. Evaporations rates of water, cosolvents, amines
   D. Homogeneity
      1. Color compatability with tinting colors
      2. Touch-up color float
   E. Wet adhesion
   F. Stain removal, washability, scrub resistance

III. Colloidal Properties in General
   A. Surface tension vs. substrate
   B. Hydrophilic/lyophilic balance (HLB)
   C. Solvent partition in polymer components (W/O or O/W)
   D. Polymer species and size distribution via gel permeations, chromatography (GPC), etc.
   E. Pigment/polymer wetting

IV. Polymer Properties—Analytical and Structural
   A. Reactive groups—carboxyls, hydroxyls, etc.
   B. Amine choice
   C. Alkyd design
   D. Polymer structure

See also Chapter 3

Architectural Paints and Coatings

The ASTM have designated specific tests for architectural paints. This standard is issued under the fixed designation D2833; the number immediately following the designation indicates the year of the original adaption or, in the case of revision, the year of last revision. A number in parentheses indicates the year of last reapproval. The procedures, methods and specifications listed in this index are classified into three groups and classes:

I. Water or Solvent Coatings
II. Waterborne Coatings
III. Solvent Coatings
   A. Interior and exterior
   B. Interior only
   C. Exterior only

See Table 19.2.

Industrial Paint Products

Tests run on industrial paint products are more varied than those for trade sales products, since the end uses are quite different.

However, most applications are to metal surfaces; so the tests emphasize such properties as the adhesion of the paint to the metal surface even after distortion of the surface due to impact, and the protection of the metal surface in salt spray and humidity testing.

Salt Spray Resistance

Salt spray tests reflect the relative ability of coatings to withstand corrosive action. The salt spray exposure test is conducted in a large chamber. Salt

TABLE 19.2. ASTM D 2833, INDEX OF METHODS.

Category		ASTM Designation	Federal Standard No. 141a
	LIQUIDS		
IA	Definitions of Terms Relating to Paints, Varnishes, Lacquers and Related Products.	D 16	Section 8
	Examination of Paint Products		
IA	Primers and Primer Surfacers over Preformed Metal	D 3322	
IIA	Resistance of Emulsion Paints in the Container to Attack by Microorganisms	D 2574	
IIA	Water Thinned Floor Paints	D 3358	
IIB	Latex Flat Wall Paints	D 2931	
IIC	Testing Asphalt Emulsions for Use as Protective Coatings for Metal	D 1010	
IIC	Exterior Latex House Paints	D 3129	
IIIA	Testing Varnishes	D 154	
IIIA	Testing Clear Lacquers and Lacquer Enamels	D 333	
IIIA	Solvent-Thinned Floor Paints	D 3383	
IIIB	Performance Tests of Clear Floor Sealers	D 1546	
IIIB	Sampling and Testing Shellac Varnish	D 1650	
IIIB	Interior Solvent-Thinned Flat Wall Paints	D 3323	
IIIB	Solvent-Based Interior Semigloss Wall and Trim Enamels	D 3425	
IIIC	Exterior Solvent-Thinned House and Trim Paints	D 2932	
	Physical Tests		
IA	Density of Paint, Varnish, Lacquer and Related Products	D 1475	4184.1
IA	Consistency of Paints Using the Stormer Viscometer	D 562	4281
IA	Evaluating the Degree of Settling of Traffic Paint	D 869	4208
IA	Leveling Characteristics of Paints by Draw-Down Methods	D 2801	
IA	Viscosity of Paints, Varnishes and Lacquers by Ford Viscosity Cup	D 1200	4282
IA	Rheological Properties of Non-Newtonian Materials	D 2196	
IA	Coarse Particles in Pigments, Pastes, and Paints	D 185	4091
IA	Fineness of Dispersion of Pigment-Vehicle Systems	D 1210	4411.1

TABLE 19.2. (Continued)

Category		ASTM Designation	Federal Standard No. 141a
IIA	Package Stability of Latex Paint	D 1849	3018
IIA	Freeze Thaw Resistance of Latex and Emulsion Paints	D 2243	3012 or 4144
IIA	Freeze Thaw Stability of Multicolor Lacquers	D 2337	
IIA	Particle Size of Multicolor Lacquers	D 2338	
IIIA	Flash Point by Tab Closed Tester	D 56	4291
IIIA	Flash Point by Pensky-Martens Closed Tester	D 93	4293
IIIA	Flash Point of Liquids by Tag Open Cup Apparatus	D 1310	
IIIA	Flash Point of Liquids by Seta-Flash Closed Tester	D 3278	
IIIA	Specific Gravity of Drying Oils, Varnishes, Resins, and Related Materials at 25/25°C	D 1963	
IIIA	Color of Transparent Liquids (Gardner Color Scale)	D 1544	4248
IIIA	Viscosity of Transparent Liquids by Bubble-Time Method	D 1545	4271
IIIA	Flow Ratings of Organic Coatings Using the Shell Flow Comparator	D 2353	
	Panel Specifications and Preparation		
IA	Preparation of Zinc-Coated Steel Surfaces for Painting	D 2092	
IA	Preparation of Hot-Dipped Nonpassivated Galvanized Steel for Testing Paint	D 2201	
IC	Wood to Be Used in Panels in Weathering Tests of Paints and Varnishes	D 358	2031
IC	Making and Preparing Concrete and Masonry Panels for Testing Paint Finishes	D 1734	2051
IIC	Asbestos-Cement Shingle Blanks to Be Used as Panels in Weathering Tests of Latex and Emulsion Paints	D 1911	
IIIA	Preparation of Steel Panels for Testing Paint, Varnish, Lacquer, and Related Products	D 609	2011.1
IIIA	Pictorial Surface Preparation Standards for Painting Steel Surfaces	D 2200	
IIIA	Preparation of Aluminum Alloy Panels for Testing Paint, Varnish, Lacquer and Related Products	D 1733	
IIIA	Preparation of Aluminum and Aluminum-Alloy Surfaces for Painting	D 1730	2013.1

TABLE 19.2. (Continued)

Category		ASTM Designation	Federal Standard No. 141a
IIIA	Preparation of Hot Dip Aluminum Surfaces for Painting	D 1731	
	Film Preparation and Thickness		
IA	Producing Films of Uniform Thickness of Paint, Varnish, Lacquer, and Related Products on Test Panels	D 823	2121 or 2162
IA	Measurement of Wet Film Thickness of Paint, Varnish, Lacquer, and Related Products	D 1212	
IA	Measurement of Dry Film Thickness of Organic Coatings	D 1005	6183
IA	Measurement of Dry Film Thickness of Nonmagnetic Organic Coatings Applied on a Magnetic Base	D 1186	6181
IA	Measurement of Dry Film Thickness of Nonmetallic Coatings of Paint, Varnish, Lacquer, and Related Products Applied on a Nonmagnetic Metal Base	D 1400	
IA	Microscopic Measurement of Dry Film Thickness of Coatings on Wood Products	D 2691	
	Drying and Curing		
IA	Drying, Curing or Film Formation of Organic Coatings at Room Temperature	D 1640	4061.1
IIIA	Gas Checking and Draft Test of Varnish Films	D 1643	4161, 4162 or 4171
	DRY FILM PROPERTIES		
	Porosity and Permeability		
IA	Moisture Vapor Permeability of Organic Coatings Films	D 1653	6171
IA	Porosity of Paint Films	D 3258	
IIA	Evaluating the Ability of Latex Paint to Resist Efflorescence from the Substrate	D 2831	
	Appearance Properties Gloss, Reflectance and Hiding Power		
IA	Specular Gloss	D 523	6101, 6103, or 6104
IA	45-deg 0-deg Directional Reflectance of Opaque Specimens by Filter Photometry	E 97	6121
IA	Hiding Power of Paint	D 2805	

TABLE 19.2. (Continued)

Category		ASTM Designation	Federal Standard No. 141a
IA	Relative Dry Hiding Power of Paints	D 344	
IA	Recommended Practice for Spectophotometry and Description of Color in CIE 1931 System	E 308	4251
IA	Specifying Color by the Munsell System	D 1535	
IA	Visual Evaluation of Color Differences of Opaque Materials	D 1729	4249.1
IA	Instrumental Evaluation of Color Difference of Opaque Materials	D 2244	6123
IA	Selecting and Defining Color and Gloss Tolerances of Opaque Materials and for Evaluating Conformance	D 3134	
	Physical Strengths and Resistance (Nonchemical)		
IA	Elongation and Tensile Strength of Free Films of Paint, Varnish, Lacquer, and Related Products with a Tensile Testing Apparatus	D 2370	6224.1
IA	Resistance of Organic Coatings to the Effects of Rapid Deformation	D 2794	
IA	Measuring Adhesion by Tape Test	D 3359	
IIIA	Elongation of Attached Organic Coatings with Conical Mandrel Apparatus	D 522	6222
IIIA	Elongation of Attached Organic Coatings with Cylindrical Mandrel Apparatus	D 1737	6221 or 6222
IIIA	Elasticity or Toughness of Varnishes	D 1642	4151 or 4152
IIIA	Print Resistance of Lacquers	D 2091	6211
IIIA	Abrasion Resistance of Coatings of Paint, Varnish, Lacquer, and Related Products with Air Blast Abrasion Tester	D 658	
IIIA	Abrasion Resistance of Coatings of Paint, Varnish, Lacquer, and Related Products by the Falling Sand Method	D 968	
IIIA	Abrasion Resistance of Clear Floor Coatings	D 1395	
IIIA	Indentation Hardness of Organic Coatings	D 1474	6212
IIIA	Adhesion of Organic Coatings	D 2197	6303.1 or 6302.1
	Resistance to Chemicals and Environment		
IA	Color Change of White Architectural Enamels	D 1543	
IA	Effect of Household Chemicals on Clear and Pigmented Organic Finishes	D 1308	6081

TABLE 19.2. (Continued)

Category		ASTM Designation	Federal Standard No. 141a
IA	Fire Retardancy of Paints (Cabinet Method)	D 1360	
IA	Fire Retardancy of Paints (Stick and Wick Method)	D 1361	
IA	Finishes on Primed Metallic Substrates for Humidity-Thermal Cycle Cracking	D 2246	
IA	Coated Metal Specimens at 100% Relative Humidity	D 2247	
IA	Coatings Designed to Be Resistant to Elevated Temperatures During Their Service Life	D 2485	
IA	Surface Burning Characteristics of Building Materials	E 84	
IA	Surface Flammability of Materials Using a Radiant Heat Energy Source	E 162	
IB	Resistance to Growth of Mold on the Surface on Interior Coatings in an Environmental Chamber	D 3273	
IB	Washability Properties of Interior Architectural Coatings	D 3450	
IB	Humid Dry Cycling for Coatings on Wood and Wood Products	D 3459	
IC	Operating Light- and Water-Exposure Apparatus (Carbon-Arc Type) for Testing Paint, Varnish, Lacquer, and Related Subjects	D 822	
IC	Accelerated Testing of Moisture Blister Resistance of Exterior House Paint or Wood	D 2366	
IC	Coated Steel Specimens Dynamically for Resistance to Corrosion	D 2933	
IC	Evaluating Degree of Surface Disfigurement on Paint Films by Fungal Growth and Dirt Accumulation	D 3274	
IC	Determining by Exterior Exposure the Susceptibility of Paint Films to Microbiological Attack	D 3456	
IC	Operating Light- and Water-Exposure Apparatus (Unfiltered Carbon-Arc Type) for Testing Paint, Varnish, Lacquer, and Related Products Using Dew Cycle	D 3361	
IIA	Stain Removal from Multicolor Lacquers	D 2198	
IIB	Scrub Resistance of Interior Latex Flat Wall Paint	D 2486	6142
IIB	Efflorescence of Interior Latex Paints	D 1736	

TABLE 19.2. (Continued)

Category		ASTM Designation	Federal Standard No. 141a
IIIA	Water Immersion Test of Organic Coatings on Steel	D 870	
IIIA	Light Stability of Clear Coatings	D 2620	
IIIA	Water Fog Testing of Organic Coatings	D 1735	
IIIA	Salt Spray (Fog) Testing	B 117	6061
IIIA	Acetic Acid Salt Spray (Fog) Testing	B 287	
IIIA	Temperature Change Resistance of Clear Nitrocellulose Lacquer Films Applied to Wood	D 1211	
IIIA	Resistance to Water and Alkali of Dried Film of Varnishes	D 1647	
IIIA	Detergent Resistance of Organic Coatings	D 2248	
IIIC	Evaluation of Painted or Coated Specimens Subjected to Corrosive Environment	D 1654	
	Durability Tests		
IA	Evaluation Degree of Blistering of Paints	D 714	6161
IC	Durability and Compatibility of Factory-Primed Wood Products with Representative Finish Coats	D 2830	
IC	Evaluating Degree of Chalking of Exterior Paints	D 659	6411
IC	Evaluating Degree of Checking of Exterior Paints	D 660	6421
IC	Evaluating Degree of Cracking of Exterior Paints	D 661	6471
IC	Evaluating Degree of Erosion of Exterior Paints	D 662	6431
IC	Evaluating Degree of Flaking (Scaling) of Exterior Paints	D 772	6441
IC	Conducting Exterior Tests of Paints on Wood	D 1006	6161.1
IC	Single- and Multipanel Forms for Recording Results of Exposure Tests of Paints	D 1150	
IIC	Reporting Paint Film Failures Characteristic of Exterior Latex Paints	D 1848	
IIIA	Evaluating Degree of Rusting on Painted Steel Surfaces	D 610	6451
IIIC	Conducting Exterior Exposure Tests of Paints on Steel	D 1014	6160
IIIC	Exterior Durability of Varnishes	D 1641	

TABLE 19.2. (Continued)

Category		ASTM Designation	Federal Standard No. 141a
METHODS FROM U.S. FEDERAL TEST METHOD STANDARD NO. 141a			
IA	Sampling (General)		1021
IA	Condition in Container		3011.1
IA	Storage Stability at Thermal Extremes		3019
IA	Reducibility and Dilution Stability		4203.1
IA	Absorption Test		4421
	Panel Specifications and Preparations		
IB	Preparation of Gypsum Wallboard Panels		2081
IIIA	Preparation of Tin Panels		2012.1
IIIA	Preparation of Glass Panels		2021
	Film Preparation		
IA	Application of Sprayed Films		2131
IA	Application of Brushed Films		2141.1
IA	Application of Film with Film Applicator (Magnetic Chuck)		2161
IA	Brushing Properties		4321.1
IA	Sag Resistance (Baker Method)		4493
IA	Sag Test (Multinotch Blade)		4494
IA	Working Properties and Appearance of Dried Film		4541
IB	Application by Rollers		2112
IB	Roller Coating Properties		4335
IIIA	Spraying Properties		4331.1
	Appearance Properties		
IA	Hiding Power (Contrast Ratio)		4122.1
IA	Color of Pigmented Coatings		4250
IA	Color Specification from Photoelectric Tristimulus Data		4252
IA	Lightness Index Difference		6122
IB	Primer Absorption and Top Coat Holdout		6261
IIIB	Accelerated Yellowness		6132
	Physical Strengths and Resistances		
IIIA	Sanding Characteristics		6321
	Resistances to Chemicals and the Environment		
IA	Mildew Resistance		6271.1

solution consisting of 5% sodium chloride in distilled water is sprayed by air pressure into the chamber through atomizing nozzles. The salt fog fills the chamber and completely envelops the test specimens, which are supported near the top of the cabinet at an angle ranging from 15 to 30° from the vertical. These specimens are small steel panels coated on one side with the material to be tested. The coating is scribed to the base metal with a cross mark, and the back and edges of the panel are protected with a coating system known to be stable under the conditions of exposure.

The salt solution is sprayed into the cabinet at 95°F. Provision is made to prevent droplets of this solution which have condensed on the top of the cabinet from falling directly on the test panels. The relative humidity during exposure approaches 100%. The panels are examined for corrosion, loss of gloss, blistering, and other effects, usually after 25, 50, 100, 250, and 500 hours. This procedure is outlined in ASTM B117.

Humidity Resistance

In humidity testing, panels are set in racks 15 to 30° from the vertical. The temperature within the cabinet is held at 100°F. The relative humidity is kept at essentially 100%. Test panels are inspected for signs of actual or imminent failure after 25, 50, 100, 250, and 500 hours' exposure in the cabinet. Failure is usually indicated by the formation of small blisters and rusting. This procedure is outlined in ASTM D2247.

COMMERCIAL TESTING LABORATORIES

Paint manufacturers and paint users have need, at times, for the services of a commercial testing laboratory, either to make analyses or tests that they are not equipped to do, or to serve as a referee.

In case you need such services and are not acquainted with a suitable commercial laboratory, The American Society for Testing and Materials, 1916 Race Street, Philadelphia, Pennsylvania 19103, publishes a *Directory of Testing Laboratories—Commercial, Institutional,* which lists laboratories by geographic location, their classification of commodities, and their measurement capabilities.

REFERENCES

Paint Testing Manual—Physical & Chemical Examination of Paints, Varnishes, Lacquers & Color, Gardner/Sward, ASTM, Thirteenth Edition, 1972.

1978 Annual Book of ASTM Standards, Part 27, "Paint—Tests for Formulated Products," Part 28, "Paint—Pigments, Resins & Polymers," and Part 29, "Paint—Fatty Oils & Acids, Solvents, Misc. Aromatic Hydrocarbons Naval Stores," ASTM.

20

Surface Preparation and Application

Surface preparation is fundamental to good painting practice. If it is done correctly, proper adhesion is obtained, and the coating system gives its maximum performance.

If old poorly adhering paint, corrosion products, or other surface contamination is not removed, failure of the coating may result in one or more of these ways:

1. There may be simple mechanical interference with proper adhesion between the coating and substrate.
2. There may be absorption of moisture or corrosive chemicals by osmosis or wicking, causing substrate corrosion.
3. Foreign matter, substrate flaws, or breaks in the coating may provide conditions for electrochemical corrosion of metal surfaces.

Substrates such as wood, masonry, metal, and plastics differ from each other in surface preparation and coating requirements. Oily and greasy surfaces are not too much of a problem with many solvent paints; however, with waterborne paints these contaminants must be removed from the surface.

SURFACE PREPARATION

Surface preparation may include one or more of the following steps:

1. Removal of loosely adhering foreign matter, corrosion products, or old paint, usually by hand methods.
2. Removal of tightly adhering foreign matter, which may later loosen and destroy adhesion or which may serve as corrosion sites. Power equipment or chemical treatment is usually required.
3. Creating surface roughness by etching, and so on.
4. Modification of the surface by chemical treatment to make it more compatible with the coating.

Hand Cleaning

The usual purpose of hand cleaning is to remove loosely adhering dirt, grease, oil, tar, rust, old paint, chemical fallout, or other foreign matter with a minimum of time and expense. Tools for hand cleaning include bristle brushes, wire brushes, scrapers, sand paper, emery paper, steel wool, and so forth.

Hand washing with water or scrubbing under the stream from a low-pressure water hose is a suitable means for removing loosely adhering foreign matter. Cleaning agents may be added to the water, but the surface must be rinsed with water to remove water-soluble residues. Steam at pressures from 200 to 1000 psi may be used.

Power Cleaning

Power cleaning may be used to remove foreign material from the surface and to roughen the surface to improve adhesion.

Power tools of many types are available and include various wire brushes, discs, hammers, and scalers. Water blasting involves the use of a water stream at 3000 to 7000 psi, supplied by a high-pressure pump.

Abrasive Blasting

Directing abrasive particles against a surface at a high velocity is one of the most effective means of surface preparation. Many studies have demonstrated that application to a blasted surface greatly prolongs the life of maintenance coatings.

Chemical Cleaning

Chemical cleaning includes acid pickling, alkali cleaning, acid cleaning, use of paint strippers, emulsion cleaning, and solvent cleaning. Chemical cleaning may be used where mechanical cleaning might damage the surface.

In pickling, an acid solution is used to dissolve or loosen rust, mill scale, and so on. Hot, strong solutions are used, and time and temperature are controlled accurately. Sulfuric, hydrochloric, and phosphoric acid solutions are most commonly used, but nitric, hydrofluoric, or organic acids are sometimes specified. Chemical cleaning has the advantage that the solution can penetrate to inaccessible spots which could not be reached by an abrasive blast. Oil and grease interfere with pickling action and must be removed beforehand, usually by solvent or alkali cleaning.

Alkaline cleaners are economical and easy to use, but require thorough rinsing. Alkaline cleaners are most effectively used on agitated soaking tanks or electrolytic cleaning tanks, or with pressure or steam spray equipment. Temperatures of 160 to 200°F are common.

Paint strippers may be alkaline, acid, or solvent-based. Application may be flow-on, immersion, or with high-pressure steam or water equipment.

Emulsion cleaners consist of a solvent such as kerosene and an emulsifier dispersed in water. Effectiveness may be increased by the addition of a mild alkali.

Solvent cleaning is effective when the soil is primarily oil, grease, fat, or wax which cannot be removed by water systems. It may be used for preliminary cleaning (i.e., before sandblasting or acid pickling), or as the final step before phosphating or painting. Methods of use include hand wiping, immersion, spray, or vapor degreasing. Immersion cleaning is done in dip tanks, which may be operated hot or cold. The cleaning tanks should be followed by one or more rinse tanks to reduce the contamination from dirty solvent. Vapor degreasing is now a widely used method of solvent cleaning. The solvents used in vapor-degreasing units should have high vapor density, stability, low specific heat, low baking point, low toxicity, and nonflammability. Chlorinated solvents such as trichloroethylene, perchloroethylene, and methylene chloride are used.

Nonmetallic Surfaces

Wood

New wood surfaces should have holes and cracks filled with putty or plastic patch. Seal knots and pitch streaks with shellac or knot sealer, sand as needed, and brush off dirt. On previously painted surfaces remove loose paint with wire brushes, scrapers, and so on. Blowtorch and propane heaters are sometimes used to remove paint from blistered sections. The paint is heated and then scraped off. Dirt, chemicals, and paint chalk must be removed by washing with water or a cleaning aid such as trisodium phosphate followed by rinsing. If there is any under-eave paint peeling, this may be caused by soluble zinc salts, which should be removed by thorough washing.

Mold or mildew must be removed, or it will survive painting and appear on the new finish. A mold-control solution should be used containing one-third cup of detergent and one-half cup of bleach per gallon of water.

Masonry and Concrete Surfaces

With these types of surfaces all loose materials should be removed by wire brushing or abrasive blasting.

APPLICATION

The paint user is concerned with processes for applying a uniform coating, usually no more than a few thousandths of an inch in thickness. Viscosity of the liquid paint is one of the most important coating properties with respect to satisfactory application to a surface.

Rheology

Rheology is the science that deals with the deformation and flow of matter. In the application of coatings, deformation and flow continue after the paint is deposited on the surface and levels. Viscosity or consistency is sometimes defined as resistance to flow. The viscosity of coatings is very sensitive to changes in temperature; it decreases as the temperature increases.

Viscosity absolute dynamic is defined as the force per unit area that resists the flow of two parallel fluid layers past one another when their differential velocity is 1 cm/(s)/cm separation.

Newton's equation is:

$$f = \eta A \left(\frac{dv}{dx}\right)$$

where A = area in cm^2, dv/dx = the velocity gradient (S′), and η = coefficient of absolute viscosity (poise). Any liquid whose coefficient of voscosity satisfies the equation, rearranged from above,

$$\eta = \frac{f/A}{dv/dx} = \frac{\text{shear stress}}{\text{shear rate}}$$

is known as a Newtonian liquid. Few liquids exhibit Newtonian or near-Newtonian behavior. Newtonian liquids include water, solvents, mineral oils, and a few resin solutions.

Most paints are non-Newtonian liquids. Non-Newtonian flow may be plastic flow, pseudoplastic flow, or dilatant flow. (See Figure 20.1.)

Plastic flow occurs when a yield stress must be overcome or exceeded before flow will take place:

$$U = \frac{\text{shear stress} - \text{yield stress}}{\text{shear rate}}$$

Pseudoplastic flow has no yield value, and the curve of the plot of shear stress vs. shear rate is nonlinear, with the shear rate increasing faster than the shear stress.

Dilatant flow is characterized by an increase in viscosity as shear stress is increased.

Thixotropy. Thixotropy is the property of a liquid to lose viscosity under

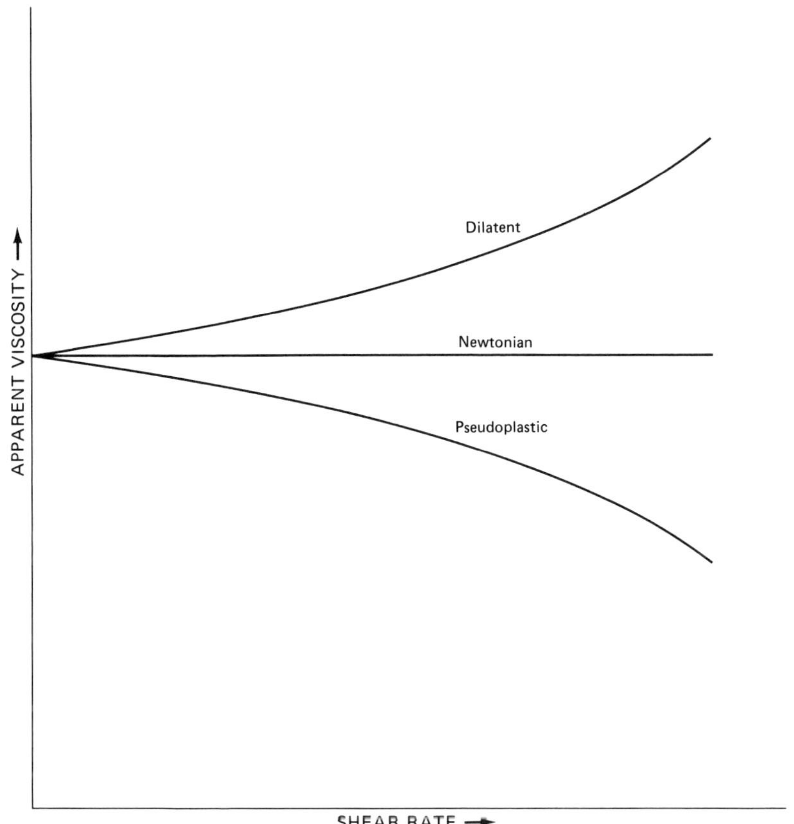

FIGURE 20.1. Viscosity behavior.

stress. Thixotropic materials are those whose consistency depends on the duration of shear as well as the rate of shear. Thixotropy is a reversible process, and after resting the structure of the material builds up gradually.

See Figure 20.2 for an example of thixotropic flow. This type of flow is frequently encountered in latex paints.

Brushing. The easiest-brushing paints generally have very pseudoplastic flow curves. Brushing viscosities for trade sales products range from 0.6 to 3.0 poises.

Spraying. Most coatings intended for spray application are reduced to relatively low viscosities and have almost Newtonian flow behavior. Typical spraying viscosities are in the range of 0.4 to 1.5 poises.

Dipping and Flow Coating. Viscosities used for dipping are about the same as spraying, 0.4 to 1.5 poises. Flow coating viscosities tend to be slightly higher than dipping viscosities.

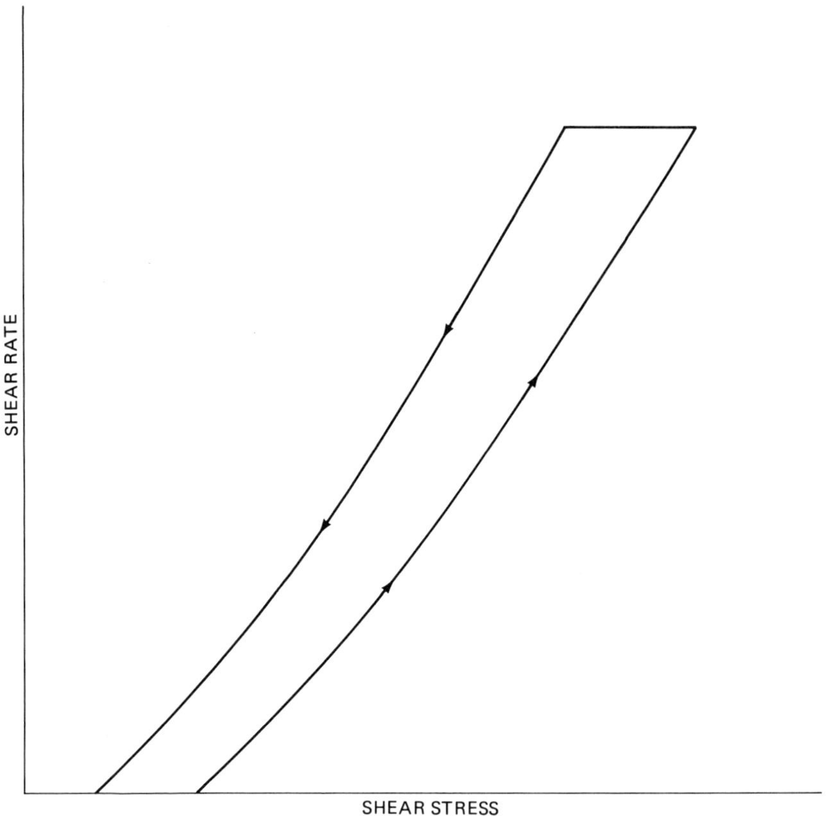

FIGURE 20.2. Thixotropy.

Type of Application

The manufacturer who is planning to modify his coating process to meet environmental regulations must consider the efficiency of his application process. The more efficient the application method is, the less paint used, and therefore the less solvent released to the atmosphere.

The Continuous Coater is one application method that approaches 100% efficiency. Fixed or reciprocating guns mounted in a closed cabinet continuously spray pieces that pass through on a conveyor. There is no exhaust on the box itself. The cabinet has openings only for product entry and exit. High-energy or venture-type scrubbers at the portals remove particulate contamination from any escaping vapors. Within the box the guns produce overspray; however, being enclosed in a saturated atmosphere, the overspray does not dry but stays liquid, falling to the bottom of the compartment where it is drawn off by a pump, passed through an automatic viscosity control, and recycled back to the spray gun. Coaters can process high volumes of constant-section workpieces; however, they are a one-color device, as it is not possible to change colors easily.

Table 20.1 shows the application efficiency of various application methods.

Brush Application

Applying paint by brush is one of the oldest techniques, but it is still used for a substantial amount of the painting done in the field. While slower than spraying, it gives excellent results when done with skill, using the proper brush and technique. Brushing works paint into pores and crevices and is the preferred way of applying primers, especially on rough surfaces.

The selection of the proper size and type of brush is essential for good work. The best brushes and the most expensive are those made of Chinese hog bristle, which, because of their split or flag-end bristles, hold an adequate amount of paint. Good nylon brushes with flag ends do a good job with waterborne paints and will outlast hog bristles four to one. The coverage per day per worker for brush application averages 1000 sq ft.

Roller Application

Roller application is a fast and convenient way of applying paint to large, flat areas that are free of protrusions. It gives excellent results on surfaces such as brick, cinder block, or plaster. It is one of the best methods for painting wire fences. Since the roller is relatively easy to use, do-it-yourselfers as well as professional painters can achieve good results. Rollers give a stippled effect.

The proper type of roller must be selected, depending on the type of paint and the applications. For waterborne paints a Dynel (acrylic) or a Dacron (polyester) roller cover is used. The rougher the surface, the longer the nap. For enamels a shorter-nap-length roller is used. The type of roller cover deter-

TABLE 20.1. APPLICATION EFFICIENCY, %.

Conventional spray:	
Air atomized	35–50
Airless atomized	40–65
Electrostatic spray:	
Air atomized	60–70
Airless atomized	45–80
Centrifugal atomized	80–90
Dip, flow, and curtain coat	70–80
Coil and roller coat	85–90
Electrodeposition	90–98

mines the leveling effect achieved. Coverage per day per worker with a roller is 2000 to 4000 sq ft. Paint for roller application should have a minimum of spatter.

Flat Pad Application

This is one of the newest methods of application of paints, and for certain applications works well. Pad applicators vary in size from 1″ × 2″ units for molding to 4″ × 7″ units for large surfaces. They consist of a foam base covered with a napped fabric. The foam serves as a reservoir. This type of application works well for stain application and floor finishing. Coverage per day per worker can be 3000 to 6000 sq ft.

Spraying

Paint spraying can be divided into two general types—conventional spray and airless.

In conventional spray the atomization is accomplished by air pressure. Compressed air is the source of energy. As compressed air is directed through the air cap, it is aimed at the material stream to form the droplets into a pattern, which is then directed onto the surface while the gun is moved. Not all paints atomize alike. The viscosity of the material is one factor that determines the atomizing pressure required. Heavy and viscous paints need high pressure, whereas low-viscosity paints will break up with low air pressure. The air pressure also determines the amount of paint flow from the gun; for example, with one particular enamel to atomize 32 oz/min required 70 pounds of air pressure, but to atomize 16 oz/min required only 45 pounds of pressure.

The energy in airless atomization is hydraulic pressure. Pressures from 1000 to 3000 psi are used. When the paint is pumped to high pressure and then directed through a small orifice on the cap of the gun, it comes out of that orifice at extremely high velocity, which breaks up the paint into atomized particles. The design and angle of the orifice are determinants of pattern size.

Conventional Spray

The chief disadvantages of conventional spraying methods are those associated with overspray caused by air and the finely divided paint bouncing off the surface being painted. The overspray wastes paint and may contaminate surrounding areas.

The simplest conventional system is the suction type. This setup consists of a gun with a suction air cup whose design produces a vacuum action around the center hole. The attached cup containing the paint has a vent hold in the lid. When air rushes through the cap center hole with the fluid tip centered there, the vacuum siphons the paint from the cup. A suction-cup unit is limited to spraying low-viscosity paints, and the amount of paint delivered is small.

For high-speed spraying, a pressure feed is used. The paint is forced to the gun through a hose from the pressure tank and gun. Pressure feed allows the use of heavy-viscosity paints and increases the speed at which the paint can be applied. The energy required in conventional air spray is about 93 hp to deliver a gallon per minute. There are approximately 600 parts of air per 1 part of paint. If the gun is too far from the surface, misting may occur, which means the paint is hitting the surface in too dry a state. Coverage per day per worker by conventional air spray may be 4000 to 8000 sq ft.

Heated Air Shroud Spray Gun

Because of high humidity conditions in spray booths, it is usually necessary to air-condition topcoat booths to keep the humidity at 50% or less to prevent sagging. One method now being tried is to use a heated air shroud spray gun to eliminate the need for air-conditioning. An attachment called a hot air shroud envelops the spray gun fan in a cowl of slow-moving hot air. Coatings sprayed under high-humidity conditions show less sagging using this equipment. Disadvantages at present are a high noise level and the fact that shroud guns are awkward to handle.

Airless Spray

Airless spray is the fastest-growing method of application. It is particularly attractive for professional painters in maintenance painting. It is a fast method of applying relatively thick films by multiple passes of the gun so that one-coat operations are possible. Reduced scaffolding results. The directional control possible with airless spraying results in more effective coverage and less loss of paint than with air spraying. Overspray misting and fogging are reduced more than 50%. Less masking of surfaces is required. Higher-solids paints may be used.

In airless spraying the pressure at the nozzle should be about 2000 psi, which for most equipment means the line pressure should be about 80 psi. If the pressure is too low, a pulsating spray, resulting in uneven film build, will occur.

The size of the orifice of the spray tip controls the amount of paint delivered. The orifice size also depends on the particle size of the pigments in the formulation. If the orifice is too small in relation to the pigment particles, clogging will occur. Excessive film build and sagging will result from using too large an orifice. A spray tip with a 15-mil orifice is suggested for initial evaluation.

In a waterborne paint a viscosity of about 60–70 Krebs units is suggested as a starting point. If the viscosity is too low, foaming will occur; if the viscosity is too high, excessive pressure will be required to force the paint through the nozzle. Some paints formulated for brush application may require 10 to 15% reduction.

Recommended spraying distance is about 15 inches. Paint should be filtered to prevent clogging.

Since these are high-pressure units, certain safety precautions must be followed:

1. Never direct the paint toward fingers, hand, etc. Under normal operating conditions there is sufficient pressure to force the paint through the skin like a hypodermic syringe. This could result in the loss of a finger, etc.

2. Airless spray units should have a pressure release valve in case of clogging in the gun or line.

Airless spray application requires about 1.1 hp of energy per gallon of paint delivered. Coverage per day is about 8,000 to 12,000 sq ft. per worker.

Airless spray is not used much on the production line because a large amount of water must be evaporated, which is a difficult undertaking, especially on air-dried coatings in confined areas.

Dipping

Dip coating is a fast and economical method of industrial finishing. Dipping is advantageous for irregularly shaped articles which are difficult to coat by other methods. This process can be carried out on a continuous conveyor operation from cleaning the part to painting and finally baking. However, while the process is simple, all variables in the operation must be closely controlled to obtain a satisfactory paint job.

Since a coating in an open tank is subject to evaporation loss and possible air oxidation, the opening should be as small as the size of the articles to be dipped will permit. There should be provision for covering the tank when it is not in use. The paint should be agitated to prevent settling.

The viscosity of the paint must be controlled carefully, as it affects the film thickness, drainage, and edge coverage, and prevents fat edges and drips in visible areas. The rate of withdrawal of the object is important; usually the faster the withdrawal, the thicker the paint film and the less the uniformity. The tank temperature should be closely controlled. Samples of the paint from the tank should be checked regularly for solids. Cosolvent may be lost more rapidly than water, so that additional solvent may have to be added with new paint. When the tank is operated on a long-term basis with the repeated additions of new lots of paint, samples should be taken periodically for accelerated stability tests in order to anticipate any possible troubles.

Flow Coating

Flow coating is a process in which the coating material is flowed onto the object from a hose, after which the excess is allowed to drain off. Flow coating is very similar to dip coating in its rquirements for close control of the coating composition, viscosity, drying rate, and drainage characteristics, but it has the advantage of not requiring a large dip tank. The coating material is supplied from a central reservoir, which can feed several hoses at a time. In both dipping and flow-coating operations the manner in which the object is hung is very important in order that it achieve a uniform coating without unsightly dry beads or fat edges. With waterborne coatings atmospheric humidity is more critical than with solvent coatings. Many times small parts are dipped and flow-coated and then centrifuged to remove excess paint.

Knife Coating

Knife coating is used for the coating of rigid flat sheets. Only one side can be coated at a time. The coating material is flowed on and spread evenly over the surface as the sheet is drawn beneath the knife or doctor blade. The coating thickness will vary with the nonuniformity in the thickness of the sheet stock.

Machine Roller Coating

This method is widely used for painting large flat sheets of metal paper, wallboard, and so on, and aluminum or steel coil stock. Rubber or steel rollers are used, with the roller rotating in a direction opposite to that of the strip being coated, while the strip is held against the coating roller by tension furnished by the driving rolls. The coating roller receives its paint from a feed roll that dips into a paint reservoir as it rotates and then offsets the paint onto the coating roll. If desired, both sides of the strip can be coated at the same time. Viscosity, line speed, and roller tolerance must be closely controlled for successful operation, but when this is done, economical, fast coating is possible. Line speeds up to 300 ft/min are used now for coating coil stock by this method.

Energy Savings

Many authorities tout the energy-saving properties of waterborne coatings, while others say there are no savings in energy. This depends on the coating system replaced and the application methods.

Some of the factors involved and their effects (comparing waterborne and solvent systems) are:

1. Latent heat of evaporation of water is greater than that of organic solvents so that more energy is required for waterborne coating.
2. The evaporation rate of water is more adversely affected by an increase

in humidity than the rates for organic solvents are, so that possibly more energy is needed for waterborne coating.

3. Less makeup air is required in ovens for waterborne systems to keep air mixture below explosive limits—less energy is needed for the waterborne systems.
4. The waterborne system usually uses a lower-temperature cure; hence it requires less energy.
5. Spray booths where waterborne coatings are used must be air-conditioned to keep humidity down to prevent sagging. This is usually required for topcoats but not for primers, and results in more energy use for waterborne systems.
6. Elimination of afterburners that may be required for solvent systems means a smaller energy requirement for waterborne systems.

Spray Booth

Spray booths for topcoats are usually air-conditioned at 75°F and 50% relative humidity as optimum conditions. Secondary spray booths for primers, and so forth, should have a thermostatically controlled temperature and humidification capable of maintaining a minimum of 30% relative humidity. One way to reduce air makeup in spray booths is to use robot spray systems. In many applications spray booth air velocity may be reduced. In one application it was lowered from 150 ft/min. to 80 ft/min.

Ovens

At the present time two basic types of ovens are used to cure industrial finishes, direct- and indirect-fired ovens.

Direct-fired ovens generally operate at 70 to 75% efficiency because much of the air is returned to the heater stage to be reused. The only heated air that is exhausted is the amount necessary to maintain an oven atmosphere at or below 25% of the LEL (lower explosion limit).

Indirect-fired ovens must heat all fresh air from ambient temperature and must expel an equal amount of air. This reduces efficiency to 45%.

Waterborne finishes have improved the efficiency of direct-fired ovens. Because of the noninflammability of water, the lower explosion limit is generally nonexistent or dramatically reduced. It is possible to recirculate more heated air and exhaust. In some cases the amount of air circulation was reduced to 25% of that previously required with a solvent system.

One rule of thumb is to exhaust 5,000 cu ft of air per gallon of water applied at the gun and 10,000 cu ft of air for every gallon of solvent applied at the gun (fire insurance regulations).

In energy utilization not all of the gross energy is available for heating purposes, because of losses of both latent and sensible heat—energy losses that generally are not considered in calculating fuel efficiency. Latent heat energy loss is that used in the chemical combustion of fuel to carbon dioxide and water. This is the energy consumed in changing water to steam. It is approximately 10% of the gross energy for gas and 7% for oil.

Sensible heat energy loss is that vented out the stack of a given system. Therefore, the higher the exhaust temperature is, the greater the energy loss. For example, theoretically at an exhaust temperature of 70°F the amount of gross energy used is 90%. At an exhaust temperature of 400°F the amount of gross energy utilized is 83%, and at 800°F it is 75%.

Savings in energy can be realized in many ways on the production line. Washers may be operated at lower temperatures without air exhaust. Air seals can be added to side-entry ovens. Coating the oven with aluminum paint reduces the gas requirements in some cases. In many cases additional insulation can be added. Using computers to control finishing system component start-up and shut-down cycles may reduce energy use over 10%.

21

Regulations Affecting the Manufacture and Use of Coatings

There are many regulations controlling the manufacture and use of coatings. These regulations are on a local, state, and national level.

Regulations relate to the following:

Metal contents	Lead, mercury and others
Solvents	Types and amounts
Flammability	Flash points of paints
Volatiles	Vinyl chloride, acrylonitrile monomers, and benzene
Asbestos	Restrictions on amount in air during manufacture
Wastes	Discharge of waste water and landfills

GOVERNMENT AGENCIES

There are currently five areas of regulatory control of major significance to the coatings industry:

1. Environmental, involving air and water quality and other factors affecting the general population.
2. Health and safety, involving the in-plant protection of workers.
3. Transport, involving the safe shipment of coatings and raw materials.
4. Coatings in direct and indirect contact with foods.
5. Consumer product safety.

The responsibility for regulating and policing these areas rests with three independent agencies operating in the executive branch of the federal government:

Environmental Protection Agency (EPA)
Consumer Products Safety Commission (CPSC)
Council on Environmental Quality (CEQ)

There are four other federal departments involved:

Department of Labor—Occupational Safety and Health Administration (OSHA)
Department of Transportation (DOT)
Department of Health, Education and Welfare (HEW)—Food and Drug Administration (FDA)
Department of Agriculture—Meat and Poultry Inspection Program

Environmental Protection Agency

EPA as established in 1970 under the control of the office of the Assistant Administrator for Toxic Substances in the executive branch. The agency was designated to centralize control over all environmental problems including air pollution from solvents and to establish standards for such pollutants. Under 1970 Amendments to the Air Quality Act, EPA also monitors implementation of state standards as required by the original legislation.

The Toxic Substance Control Agency, formed as a result of the provisions of the Toxic Substances Control Act (TSCA) (15USC2607) and administered by EPA, is currently concerned specifically with air, water, and landfill contamination control.

At present, there are no nationwide regulations governing the type and amount of solvent vapors that may be released into the atmosphere. Such restrictions are being developed on the basis of regional climate, industrial distribution, and population density.

The EPA has proposed limitations on the solvent contents of various types of coatings. These proposals are expected to be adopted by most of the states. These limits reduce the allowable volatile organics to those in waterborne or high solids coatings.

Limits are usually based on the pounds of organics per gallon of coating less water. (See Figure 21.1.) Some of the regulations, such as flat stock, now specify the pounds of volatile organics per thousand square feet of coated surface. (See Table 21.1.)

The State of California has adopted the following regulation on architectural coatings:

		Recommended emission limit (lbs of organic/gal of coatings less water)
Sep. 1979	Exterior	2.03
	Interior	2.92
Sept. 1980	Exterior	2.03
	Interior	2.03
Sept. 1982	No exemptions	

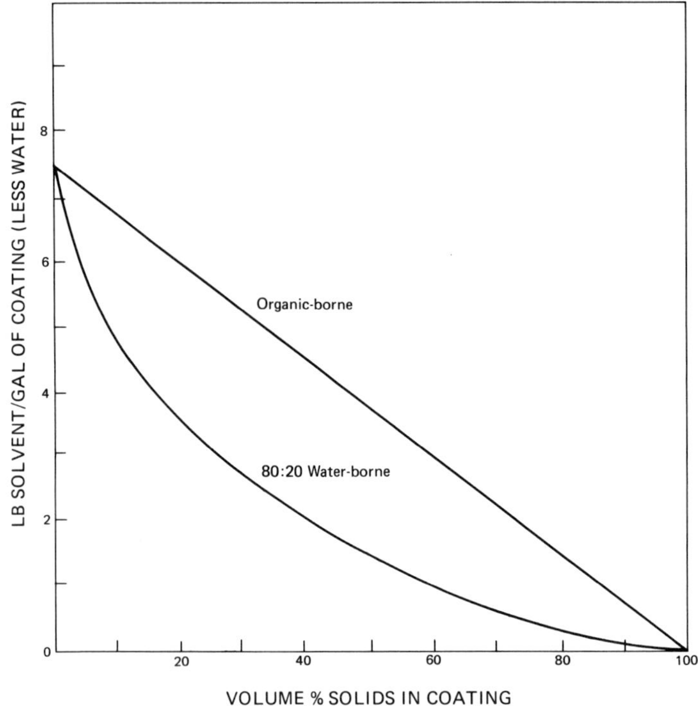

FIGURE 21.1. Weight of organic solvent per gallon as a function of solids content.

OSHA

The Occupational Safety and Health Act of 1970 established the Occupational Safety and Health Administration as well as creating an independent agency, the Occupational Safety and Health Review Commission (OSHRC). The purpose of OSHRC is to adjudicate contested enforcement actions brought by OSHA.

OSHA standards have generally been based on the recommendations of the National Institute of Occupational Safety and Health (NIOSH). This is a research organization that has no regulatory powers and operates under the control of the Department of Health, Education and Welfare's Public Health Service Center for Disease Control, which was established in 1973.

Two series of OSHA regulations are pertinent to waterborne coatings. The first series is 29CFR 1901.106, "Flammable and Combustible Liquids," which covers handling and safety aspects of solvent-containing coatings as to storage, tank venting, and equipment design.

The second regulation is 29CFR 1910.100, "Air Contaminants," which covers volatile-substance monomers (e.g., vinyl chloride, acrylonitrile) and solvents (e.g., benzene). These regulations established allowable amounts of vapors that workers may be exposed to in the work place.

Department of Transportation

DOT, successor to the Interstate Commerce Commission, which was set up in 1877, is the oldest federal regulatory agency concerned with business and

TABLE 21.1. EPA RECOMMENDED EMISSION LIMITS OF VOLATILE ORGANIC SOLVENT.

Date due: Jan. 1	Category	Recommended emission limit (lb of organic/gal of coating less water)
1979	Fabric coating	2.9
1979	Vinyl coating	3.8
1979	Paper coating	
1979	Can coating	
	Inside spray	4.2
	Base coat	2.8
	Side seam spray	5.5
	End seal compound	3.7
1979	Coil coating	2.6
1979	Auto assembly	
	Primer	1.9
	Topcoat	2.8
	Final repair	4.8
1979	Metal furniture	3.0
1979	Magnet wire coating	1.7
1979	Large appliances	2.8
1980	Miscellaneous metal parts	
	Clear coat	4.3
	Air-dried	3.5
	Outdoor parts	3.5
	Frequent color change	3.0
	Powder appliance lines	0.4
	Other	3.0
1980	Flat wood paneling	
	Printed finish interior	6.0[a]
	Natural finish	12.0[a]
	Class II finish	10.0[a]
1980	Graphic arts	Not yet available
	Architectural coatings	Not yet available
	Wood furniture manufacture	Not yet available

[a]lb of volatile organic/1000 sq ft of coated surface.

commerce. Regulations governing the interstate and, in certain cases, intrastate transportation of coatings are handled by DOT.

Consumer Protection Safety Commission

The Consumer Protection Safety Commission is an independent federal regulatory agency established in 1972. CPSC is responsible for setting standards designed to reduce the risk of injury to users of household items sold at the retail level. Areas not covered by other agencies may be regulated by CPSC.

FOOD PACKAGING

The Food and Drug Act of 1906 is among the oldest of federal regulatory provisions. At present two cabinet-level departments are involved: the Department of Health, Education and Welfare, which controls the FDA; and the Department of Agriculture, which, through its Meat and Poultry Inspection Program, controls coatings used in federally inspected meat and processing plants.

All regulations pertaining to food-packaging materials can be found in Title 21 Code of Federal Regulations Part 100 to 199, which is available from the Superintendent of Documents, Washington, D.C. Of particular interest to the paint and coatings industry is Section 175.300, "Resinous and Polymeric Coatings."

FLAMMABILITY

The CPSC administers the Federal Hazardous Substances Act (FHSA). These regulations deal with the general subjects of flammability and toxicity with the intent of reducing the potential of products for causing injury in and around the household. Under the FHSA, hazardous substances intended or packaged in a form suitable for use in the household are required to be labeled in accordance with the Act. "Hazardous substances" are defined to include flammable or toxic substances that may cause substantial personal injury or illness.

Flammability is divided into three categories, depending upon the flash point of the product (FHSA Sect 2) and the regulations at 16CFR 150.43 (method of test for flash point of volatile flammable materials by Tagliabue open cup apparatus). Volatile materials are classified as:

Extremely flammable: any substance that has a flash point at or below 20°F.
Flammable: flash point above 20°F to/and including 80°F.
Combustible: flash point above 80°F to/and including 150°F.

AIR POLLUTION

Rule 66

The first legislative control of emission of solvents, the well-known Rule 66, was enacted by the Los Angeles Air Pollution Control District in July 1966. Rule 66 was based on experimental work that demonstrated that part of the smog common in the Los Angeles basin could be the result of atmospheric reactions involving certain solvents. Researchers believed that the chemical reactions that cause smog occur when certain types of solvent vapors and in nitrogen dioxide combine by a photochemical reaction in the presence of sunlight.

A photochemically reactive solvent is considered to be any solvent with an aggregate of more than 20% of its total volume composed with an aggregate of more than 20% of its total volume composed of the chemical compounds listed below and which exceeds any of the following individual percentage composition limitations, referred to the total volume of solvent.

1. A combination of hydrocarbons, alcohols, aldehydes, esters, ethers, or ketones having an olefinic or cylco-olefinic type of unsaturation: 5%.
2. A combination of aromatic compounds with eight or more carbon atoms to the molecule, except ethylbenzene: 8%.
3. A combination of ethylbenzene, ketones having branched hydrocarbon structure, trichloroethylene, or toluene: 20%.

The provisions of this rule do not apply to materials in which the volatile material consists only of water and organic solvents, and the organic solvents comprise not more than 20% by volume of said volatile content, and the volatile content is not photochemically reactive materials as defined previously.

Many states adopted Rule 66–type regulations, but they are now being replaced with newer regulations.

About 1974 work was done that led to the development of the theory of the "Transport Phenomena." This work showed that so-called exempt solvents were carried by the prevailing winds, and that while they were nonreactive in the areas where they were released, they could be transported a distance of perhaps 300 miles and then cause irritation there.

Less reactive hydrocarbons within an urban area can persist from one day to the next under stagnant conditions and then have time to react and contribute to oxidant levels. Also, under conditions where transport of oxidants and precursors occurs, there can be sufficient time for less reactive organics to contribute to oxidant levels in neighboring areas.

EFFLUENT REGULATIONS

The paint industry uses about 80 million gallons of water per day, of which less than 5% is contaminated by virtue of its use. The EPA is now developing regulations controlling the discharge of effluent into surface waters.

METALS

Regulated metals of most importance to waterborne coatings are mercury and lead.

Mercury

Phenylmercuric compounds have been used in paints as a biocide for years, both as preservatives and as fungicides. Concern about mercury occurred about ten years ago when high concentrations of mercury were found in many lakes, resulting in a high mercury content in the fish there. Mercury is present in the earth's crust, and it has been found in lakes never visited by man. At one time it was thought that inorganic mercury in lakes could be converted by bacteria to the very toxic methyl mercury. However, tests show that the reverse is true: bacteria convert methyl mercury to inorganic mercury and methane. The EPA has not set limits on the amount of mercury to be used, and simply states it as being consistent with good practice. However, the Federal Hazardous Substance Act limits the use to 0.2% mercury (calculated as metal), based on the total weight of the paint. The current OSHA Standard (TLV) for inorganic and phenyl-mercury compounds indicates a level of 150 $\mu g/m^3$.

Lead

The Consumer Products Safety Commission has set limits on lead of 0.06% by weight, calculated as the metal on the dried paint film on consumer products paints used around the home and schools—both interior and exterior paints, toy finishes, and so on. There is an exemption on factory-applied finished to metal furniture on the basis that such finishes are inaccessible to children; that is, they cannot be separated easily from the surface and ingested by children. The Department of Housing and Urban Development is instructed to prohibit lead-based paint (0.06% lead) in residential structures built or rehabilitated by the federal government. Lead is used in many pigments, but they can now be used only in maintenance and industrial finishes. Lead is used as a drier in water-soluble coatings containing drying oils; so lead must be replaced with other metals.

Other Metals—Antimony, Arsenic, Cadmium, Selenium, Barium

The American National Standards Institute (ANSI) has set up specifications to minimize hazards to children from residual surface coating materials. The specifications state that:

A liquid coating material to be deemed suitable, from a health standpoint, for use on articles such as furniture, toys, etc., or for interior use in dwelling units where the dry film might be ingested by children:

1. Shall not contain lead compounds of which the lead content (calculated as Pb) is in excess of 1% of the contained solids (including pigments, film solids, and driers);
2. Shall not contain compounds of anitmony, arsenic, cadmium, mercury, or selenium of which metal content individually or in total (calculated as Sb, As, Cd, Hg, Se, respectively) is in excess of 0.06% by weight of the contained solids (including pigments, film solids, and driers);
3. Shall not contain barium compounds of which the watersoluble barium (calculated as Ba) is in excess of 1% of the total barium in such coatings.

22
Coating Calculations

Calculations are involved in the formulations applications, and environmental regulations pertaining to coatings.

A prime consideration in coating calculations is the importance of volume relationship of ingredients. Although raw materials are generally purchased and used on a weight basis (pounds), paints are sold on a volume basis (gallons). Usage is also on a volume basis—coverage (square feet at a specified film thickness).

DEFINITIONS

The following terms are used in paint formulation:

1. *Specific Gravity:* the ratio of the weight of a given volume of material to the weight of an equivalent volume of water, usually measured at a temperature of 77°F (25°C).
2. *Pounds per Gallon:* the weight in U.S. pounds of one gallon of a particular raw material or finished product.
3. *Bulking Value:* the volume in gallons obtained from one pound of a raw material. Mathematically it is the reciprocal of pounds per gallon.
4. *NVM–Weight:* nonvolatile material by weight of the vehicle or paint.
5. *NVM–Volume:* volumetric solids.
6. *Pigment Volume Concentration (PVC):* the percentage by volume of pigment on the total nonvolatile portion.
7. *Pigment/Binder Ratio (P/B):* the ratio of pigment to binder in the coating, commonly expressed by using unity for the P figure.

TYPICAL RAW MATERIAL PHYSICAL CONSTANTS

Physical constants of raw materials are shown in the following chart:

	Specific gravity	lb/gal	Bulking value, gal/lb
Water	1.000	8.33	0.120
Titanium dioxide (rutile)	4.20	34.99	0.02858
Titanium dioxide (anatese)	3.88	32.32	0.03094
Lithopone ($ZnS/BaSO_4$)	4.29	35.74	0.02798
Linseed oil (raw)	0.93	7.72	0.130
Mineral spirits	0.785	6.55	0.153
Toluene	0.870	7.25	0.138
Iron oxide	5.20	41.82	0.02391
Carbon black	1.81	15.08	0.06631
Acrylic latex (46% solids)	0.933	7.77	0.112

COMPOSITION

The following chart shows weight and volume relationships for a coating:

		% Wt.	% Vol.
Pigment	P	36.25	13.02
Total volatile	TV	45.45	64.73
Nonvolatile vehicle	NVV	18.30	22.25
		100.00%	100.00%
Vehicle: Vehicle solids		28.70	
Vehicle volatile		71.30	
		100.00%	

PAINT APPLICATION: COST ANALYSIS

The cost of painting a square foot of surface can be determined, but before we proceed with a cost analysis, a few terms should be defined:

1. *Spreading Rate:* the area that the coating can be spread to or cover. Units—square feet per gallon.
2. *Volume Solids:* the quantity of a liquid coating (pigment, binder, and additives) minus the volatile that evaporates. The volume solids of every coating should be given by the supplier and should not be confused with weight solids of a coating. Units—a percent (%) of the quantity in a gallon.
3. *Dry Film Thickness (DFT): the cross-section measurement of the dry coating. Units–mils, or thousandths of an inch.*

A liquid coating with 100% volume solids and applied at 1 mil dry film thickness will cover 1604 square feet per gallon.

This figure is obtained from the following formula:

$$\frac{\text{cu in./gal} \times \text{volume solids}}{\text{sq in. (sq ft)} \times \text{in.}} = \text{Spreading rate (sq ft/gal)}$$

$$\frac{(231)(1.0)}{144 \times 0.001} = 1604 \text{ sq ft/gal @ 1 mil dry film @ 100\% volume solids}$$

Using this basic formula, the spreading rate of any liquid coating is as shown in Table 22.2. This table shows the spreading rate of a coating at various values of volume solids applied at various dry film thicknesses (DFT).

For example, if the volume solids of a formula was 45% at application viscosity and a 1.5-mil dry film, the area covered by one gallon of paint would be 480 square feet:

$$\frac{231 \times 0.45}{144 \times 0.0015} = 480 \text{ sq ft/gal spreading rate}$$

The cost per gallon must be determined at the application conditions. Therefore the cost of any solvent must be included in the final cost, in addi-

TABLE 22.1. PAINT FORMULA CALCULATIONS.

The following is an example of paint formula calculations. Substitute column values for item numbers.

	a lb.	b gal.
1. Titanium dioxide	250	7.44
2. Inert—talc	180	5.58
3. Acrylic latex, 46%	472	52.86
4. Water	284	34.12
5. Total	1186	100.00

Weight solids $\frac{1a + 2a + (.46)(3a)}{5a} = 54.55\%$

Volume solids $\frac{1b + 2b + 22.25}{5b} = 35.27\%$

Pigment volume content $\frac{1b + 2b}{35.27} = 36.92$

Titanium dioxide, pounds per gallon $\frac{1a}{5b} = 2.5$ lbs.

Weight/gallon $\frac{5a}{5b} = 11.86$ lbs.

TABLE 22.2. SQUARE FEET PER GALLON SPREADING RATE AT VARIOUS DRY FILM THICKNESSES (DFT) (IN MILS).

Vol. NVM, %	0.5 mil	1.0 mil	1.5 mil	2.0 mil	2.5 mil	3.0 mil	4.0 mil
10	320	160	107	80	64	53	40
15	480	240	160	120	96	80	60
20	640	320	214	160	128	107	80
25	800	400	260	200	160	133	100
30	960	480	320	240	192	160	120
35	1120	560	370	280	224	187	140
40	1280	640	425	320	256	213	160
45	1440	720	480	360	288	240	180
50	1600	800	530	400	320	267	200
55	1760	880	586	440	350	293	220
60	1920	960	640	480	384	320	240
65	2080	1040	695	520	415	347	260
70	2240	1120	748	560	450	374	280
75	2400	1200	800	600	480	400	300
80	2560	1280	854	640	510	427	320
85	2720	1360	906	680	545	453	340
90	2880	1440	960	720	575	480	360
95	3040	1520	1010	760	608	505	380
100	3200	1600	1065	800	640	533	400

tion to determining the volume solids at application viscosity. If a paint requires 33% reduction for spraying viscosity, and the per-gallon cost of the reducer is \$1.00 (with cost of the paint unreduced \$4.00 per gallon), the final cost of the product at application viscosity is:

$$\frac{\text{Cost of gallon unreduced} + \text{Cost of reducer required}}{\text{Total gallons}} = \text{Cost/gallon}$$

or:

$$\frac{\$4.00 + \$0.33}{1.33} = \$3.26$$

The cost analysis of paint as part can be calculated from spreading rate and application efficiency:

$$\frac{\text{Cost per gallon}}{\text{Spreading rate} \times \text{Deposition efficiency}} = \text{Cost per square foot}$$

Example:

Volume solids = 45% @ application viscosity ⎤
Recommended dry film thickness = 1.5 mil ⎦ 480 sq. ft. gal. (from Table 22.2)

Conventional air spray = 50% efficiency

$$\frac{\$3.26}{480 \times .50} = \$0.0136 = \text{Cost per square foot}$$

DETERMINATION OF VOLATILE ORGANIC CONTENT IN COATINGS
(Volatile organic content in organic coatings is required to meet certain government air control regulations.)

Data required for the determination are:

Coatings density: D (lb/gal)
Paint composition: NVM, V, VOC, H_2O

where:

NVM	=	Nonvolatile matter (wt	
V	=	Volatiles including water	
VOC	=	Volatile organic compounds	7.36 lb/gal
H_2O	=	Water	8.33 lb/gal

For Conventional Paint

Data in weight %:

$$\text{VOC} = \left[\frac{\%\ \text{VOC}}{100}\right] (\text{D})\ \text{lb/gal}$$

If % solvent by weight was 50% and weight per gallon (D) of paint was 9.5, then:

$$\text{VOC} = (0.50)(9.5) = 4.75\ \text{lb solvent/gal}$$

Data in volume %:

$$\text{VOC} = \left[\frac{\%\ \text{VOC}}{100}\right] (7.36)\ \text{lb/gal}$$

If % solvent by volume is 35% and solvent weight per gallon is 7.36:

$$\text{VOC} = (0.35)(7.36) = 2.58\ \text{lb solvent/gal}$$

For Waterborne Paint

Data in weight %:

$$\% \text{ of total coating: VOC} = \left[\frac{\% \text{ VOC}}{100}\right] (D) \text{ lb/gal}$$

$$\% \text{ of volatiles: VOC} = \left[\frac{\% \text{ VOC}}{100}\right] \left[\frac{\% \text{ V}}{100}\right] (7.36) \text{ lb/gal}$$

Data in volume %:

$$\% \text{ of total coating: VOC} = \left[\frac{\% \text{ VOC}}{100}\right] (7.36) \text{ lb/gal}$$

$$\% \text{ of volatiles: VOC} = \left[\frac{\% \text{ VOC}}{100}\right] \left[\frac{\% \text{ V}}{100}\right] 7.36 \text{ lb/gal}$$

For VOC (lb/gal less water):

$$\text{VOC} = \frac{(\text{VOC lb/gal})}{\left(1 - \frac{\text{Volume } \% \text{ } H_2O}{100}\right)} \text{ lb/gal less water}$$

For a waterborne coating with 25% volume solids and 75% solvent, in which 80% of solvent is water and 20% is organic solvent:

$$\text{Volume water: } 0.80 \times 0.75 = 0.6 \text{ gallon}$$

$$\text{Volume organic solvent: } 0.75 - 0.6 = .15 \text{ gallon}$$

$$\text{VOC} = \frac{(0.15)(7.36)}{1 - 0.6} = 2.76 \text{ lb/gal}$$

Conversion factor: lb/gal times 0.12 = kg/liter.

WEIGHT OF VOLATILE ORGANIC COMPOUNDS PER 1000 SQUARE FEET OF FINISHED PRODUCT

Two important factors must be known: (1) lb VOC/gal and (2) spreading rate in ft^2/gal.

$$\text{lb VOC}/1000 \text{ ft}^2 = \frac{\text{lb VOC/gal} \times 1000}{\text{Spread rate in ft}^2/\text{gal}}$$

Example:

$$\text{lb VOC/gal of coating} = 4.20$$

$$\text{Spreading rate} = 1800 \text{ ft}^2$$

$$\text{lb VOC}/1000 \text{ ft}^2 = \frac{4.20 \text{ lb/gal} \times 1000}{1800 \text{ ft}^2 \text{ gal}}$$

$$\text{lb VOC}/1000 \text{ ft}^2 = 2.33$$

CONVERSION FACTORS

Length:

1 inch = 2.54 centimeters	1 centimeter = 0.3937 inch
1 foot = 0.3048 meter	1 meter = 3.281 feet

Dry Volume:

1 cubic foot = 0.0283 cubic meter	1 cubic meter = 35.31 cubic feet

Liquid Volume:

1 U.S. gallon = 3.785 liters	1 liter = 0.2642 gallon
1 U.S. gallon = 231 cubic inches	

Weight Conversion:

1 pound = 0.4536 kilogram	1 kilogram = 2.205 pounds

Miscellaneous:

Density:

1 gallon of water at 60°F = 8.33 lb
1 cubic foot of water at 60°F = 62.37 lb

Pressure:

1 atmosphere = 760 mm mercury or 29.92 inches mercury

GLOSSARY

Acicular Pigments. Pigments whose particles are needle-shaped.

Acid. A chemical compound containing one or more hydrogen atoms available for reaction with active metals or alkaline solutions.

Acid Number. The number of milligrams of potassium hydroxide required to neutralize the free fatty acid in one gram of fat, oil, wax, or resin.

Acrylic. A family of synthetic resins made by polymerizing esters of acrylic acids.

Adhesion. The property that causes one substance to stick to another. There are two types of adhesion that are of interest with paint. Mechanical adhesion depends upon the penetration of the paint into the pores of the surface. Molecular adhesion is observed on smooth surfaces where the polar group in the vehicle (i.e., carboxyl or hydroxyl) attaches itself to the surface.

Adsorption. The adhesion of a substance to the surface of a solid or liquid. Pollutants are extracted by adsorption on activated carbon or silica gel.

Aerobic. Term referring to processes that can occur only in the presence of oxygen.

Afterburner. An air pollution abatement device that removes undesirable organic gases through incineration.

Agglomerate. Clusters of pigment particles that settle out and have poor optical properties.

Air Pollution. The presence of contaminants in the air in concentrations that interfere directly or indirectly with man's health, safety, or comfort.

Air Quality Standards. The prescribed level of pollutants in air that cannot be exceeded during a specified time in a specified geographical area.

Airless Spray. A system of applying paint under high pressure in which the paint is broken up into droplets when it enters the lower-pressure region outside of the gun tip. Very little air is used as compared to conventional air spray.

Aliphatic Solvent. Solvent composed of straight-chain hydrocarbons; examples: kerosene, naphtha, mineral spirits.

Amorphous. Lacking crystallinity.

Antiskinning Agent. A type of antioxidant, usually volatile that, when added to a varnish or paint, will inhibit the skinning tendency.

Aromatics. Compounds containing at least one benzene ring. Benzene, toluol, and xylol are aromatic solvents.

Binder. The nonvolatile portion of the coating vehicle which is a film-forming ingredient used to bind the pigment particles together.

Biocide. An agent to destroy microorganisms.

Biodegradable. The process of rapid decomposition as a result of the action of microorganisms.

Bleeding. The phenomenon in which a pigment in the undercoat is soluble in the vehicle in the topcoat, resulting in a discoloration on the paint surface from the pigment diffusing through.

Blistering. The formation of bubbles on the painted surface caused by moisture or solvent trapped under the surface.

Bloom. A clouding condition of the paint film, usually caused by reactive materials in the paint film coming in contact with atmospheric gases, dust, or moisture.

Brushability. The ability or ease with which a paint can be brushed.

Build. The apparent thickness or depth of a paint after application and drying.

Burnishing. Shiny or lustrous spots on a paint surface caused by rubbing.

Calcimine. A water paint composed of calcium carbonate or clay and glue.

Carcinogenic. Cancer-producing.

Casein Paint. A paint in which casein is the principal binder.

Cathodic Protection. The prevention of corrosion of a metal by electrically connecting it to a sacrificial anode. The anode can be replaced, but the metal is protected.

Chalking. The presence of loose powder on the surface of paint caused by deterioration of the binder due to exposure which releases the pigment. Controlled chalking results in a self-cleaning paint.

Checking. Paint failure in which small cracks appear on the surface of the paint.

Chroma. Color intensity or purity of tone, being the degree of freedom from gray.

Coagulation. The precipitation of a colloid into a solid mass.

Coalescence. The flowing and fusing together of soft particles.

Cohesion. The attractive forces between polymers of a single nature that tend to hold the polymers together.

Colloid. A dispersion of ultramicroscopic particles of a solid, liquid, or gas in a different medium, which can be solid, liquid, or gas.

Color. The generic term referring inclusively to all of the colors of the spectrum, white, and black, and all tints, shades, and hues that may be produced by their admixture: hue, tint, and shade.

Colorfast. Fade-resistant.

Color Retention. The property that a material has when it is exposed to the elements and shows no signs of changing color.

Copolymer. The produce formed when two or more monomers are polymerized together to yield a polymer.

Cratering. Small depressions in a paint film that may or may not expose the underlying surface.

Crawling. The tendency of a wet paint film to recede from certain areas of a painted surface; often caused by surface contamination.

Detergent. Synthetic material that, like soap, lowers the surface tension of water, emulsifies oils, and holds dirt in suspension.

Dew Point. The temperature at which water condenses from the air. The dew point varies with relative humidity.

Dilatancy. The property of some materials by which the resistance to flow increases with agitation.

Dirt Collection. The accumulation of dust, dirt, and other foreign matter on a paint surface.

Dispersion. A generic term describing any heterogeneous system of solids, gases, and liquids.

Drier. A catalytic material that, when added to drying oil or resins containing drying oils (alkyds, etc.), accelerates the drying or hardening of the film. Driers are usually in the form of metal salts of lead, cobalt, manganese, etc.

Drying Oils. Oils, usually obtained from plant sources, that will harden on exposure to air. They contain three or more double bonds. Linseed oil is a common example.

Ecology. The interrelationships of living things to one another and to their environment.

Ecosystem. The interacting system of a biological community and its nonliving environment.

Efflorescence. A phenomenon whereby a white deposit of fine crystals forms on a painted surface. These crystal are usually sodium salts which diffuse through the paint surface from the substrate.

Electrodeposition. The process by which electrically charged paint is plated on conductive surfaces of the opposite charge.

Electrostatic Spraying. A system of applying paint in which the sprayed paint droplets are given an electrical charge that results in their attraction to the grounded work piece.

Emission Factor. The average amount of pollutants emitted from a polluting source per unit of material produced.

Emulsifier. A material that, when added to a mixture of dissimilar materials such as oil and water, will produce a stable homogeneous emulsion.

Emulsion. A suspension of fine particles or globules of a liquid within a liquid, which is usually water.

Emulsion Polymerization. The process of polymerization taking place in the presence of water to form a latex.

Environment. The sum of all external conditions and influences affecting the life, development, and survival of an organism.

Exempt Solvents. Solvents whose use is not subject to certain air pollution legislation. Many alcohols, many esters, some ketones, and mineral spirits are exempt under Rule 66.

Extender Pigment. A pigment that contributes very little hiding to the system but does reinforce the paint.

Flash Point. The temperature at which a coating or solvent will ignite when exposed to flame. The lower the flash point, the greater the hazard.

Floating. Separation of pigment colors on the surface of the applied paint.

Flocculation. Clumping together of pigment particles within the wet paint.

Fungi. Small, often microscopic plants without chlorophyll.

Fungicide. A chemical that kills fungi.

Glass Transition Temperature. The temperature at which polymer molecules are able to move freely even in the solid state. Rubber is elastic because the molecules at room temperature are above their glass transition temperature and can easily be stretched out of position.

Gloss. The degree to which a surface reflects light.

Grain Raising. Swelling and standing up of the wood grain caused by absorbed water or solvent.

Heat Bodied. Thickening or polymerizing a drying oil by heat processing.

Heavy Metals. Metallic elements with high molecular weights generally toxic to plant and animal life. Examples: mercury, chromium, cadium, arsenic, lead, etc.

Hiding Power. The ability of a paint to obscure the background over which it is applied.

Homopolymers. A polymer produced from a single monomer.

Hot Spray. Process in which paint is heated prior to spraying to reduce viscosity so that higher solids may be applied.

Hue. The specific quality distinguishing one color from another, such as yellow, red, blue.

Humidity. A measure of the amount of moisture in the air. Absolute humidity reflects the actual quantity of moisture in the air. Relative humidity is the percentage of moisture in the air at a specific temperature as compared to saturation.

Hydrophilic. Water-loving or attracted to water.

Hydrophobic. Water-hating or repelled by water.

Intercoat Adhesion. The adhesion between two coats of paint.

Intumescent. Foaming and swelling up as a result of heat. Some fire-retardant paints exhibit intumescence, forming a heat-insulating surface.

Latex. A generic term describing a stable dispersion of insoluble resin particles in a water system.

Lipophilic. Oil-loving.

Monomer. An organic compound capable of polymerizing or linking together with itself or other monomers.

Orange Peel. A pebbled effect on the surface of a paint film from not leveling.

Overspray. Sprayed paint that misses the area being painted and falls upon the surroundings.

Particulates. Finely divided solid or liquid particles in the air or in an emission. Particulates include dust, smoke, fumes, mist, spray, and fog.

pH. A measure of the acidity or alkalinity of a material. pH is represented on a scale of 0 to 14, with 7 representing a meutral state, 0 the most acid, and 14 the most alkaline.

Photochemical Oxidants. Secondary pollutants formed by the action of sunlight on the oxides of nitrogen and hydrocarbons in air. They are the primary contributors to photochemical smog.

Pigment. A finely divided insoluble material that imparts color or black and white and hiding to the material to which it is added.

Pollution. The presence of matter or energy whose nature, location, or quantity produces undesired environmental effects.

Polymer. A high-molecular-weight material composed of large number of repeating units of monomer linked together.

Polymerization. Reaction of single chemical compounds, monomers, to form multiple units, or polymers.

Polyvinyl Acetates. (Commonly called PVA's) Homopolymers or copolymers of vinyl acetate with more flexible monomers.

Resin. A solid or semisolid organic material of either natural or synthetic origin.

Sagging. Excessive flow in a paint during application.

Solution. A homogeneous mixture of a dissolved substance in a liquid.

Spreading Rate. The area of surface over which a unit volume of paint will spread, usually expressed in square feet per gallon.

Styrene-Butadiene. A copolymer of styrene and butadiene made by emulsion polymerization for use in latex paints.

Surfactant. A chemical compound that changes the surface forces of a liquid or solid in relation to other gases, liquids, or solids.

Surface Tension. The tension exhibited by the free surface of liquids, measured in dynes per centimeter.

Thixotropy. The property of a liquid to lose viscosity under stress.

Throwing Power. The ability of an electrodeposition resin to coat recessed areas.

Vehicle. The film-forming or pigment-binding portion of a coating.

Vinyl. The general term applied to a class of resins containing many different materials such as polyvinyl chloride, polyvinyl acetate, etc.

Viscosity. The resistance of a liquid to flow.

Wrap-Around. The phenomenon by which electrically charged paint droplets curve around to the rear side of the object being painted.

REFERENCES

Paint/Coating Dictionary, Federation of Societies for Coating Technology 1978, Philadelphia, Pa.

Index

Index